Human–Computer Interaction Series

Editors-in-Chief

Desney Tan
Microsoft Research, Redmond, WA, USA

Jean Vanderdonckt
Louvain School of Management, Université catholique de Louvain,
Louvain-La-Neuve, Belgium

The Human–Computer Interaction Series, launched in 2004, publishes books that advance the science and technology of developing systems which are effective and satisfying for people in a wide variety of contexts. Titles focus on theoretical perspectives (such as formal approaches drawn from a variety of behavioural sciences), practical approaches (such as techniques for effectively integrating user needs in system development), and social issues (such as the determinants of utility, usability and acceptability).

HCI is a multidisciplinary field and focuses on the human aspects in the development of computer technology. As technology becomes increasingly more pervasive the need to take a human-centred approach in the design and development of computer-based systems becomes ever more important.

Titles published within the Human–Computer Interaction Series are included in Thomson Reuters' Book Citation Index, The DBLP Computer Science Bibliography and The HCI Bibliography.

More information about this series at http://www.springer.com/series/6033

Rebekah Rousi · Jaana Leikas · Pertti Saariluoma
Editors

Emotions in Technology Design: From Experience to Ethics

Editors
Rebekah Rousi
Faculty of Information Technology
University of Jyväskylä
Jyväskylä, Finland

Jaana Leikas
VTT Technical Research Centre
of Finland Ltd
Tampere, Finland

Pertti Saariluoma
Faculty of Information Technology
University of Jyväskylä
Jyväskylä, Finland

ISSN 1571-5035   ISSN 2524-4477   (electronic)
Human–Computer Interaction Series
ISBN 978-3-030-53485-1   ISBN 978-3-030-53483-7   (eBook)
https://doi.org/10.1007/978-3-030-53483-7

This Springer imprint is published by the registered company Springer Nature Switzerland AG
The registered company address is: Gewerbestrasse 11, 6330 Cham, Switzerland

Acknowledgements

First and foremost, we would like to thank Helen Desmond from Springer Nature for her immense patience and strong professionalism while dealing with us and our manuscript. We cannot begin to expression our appreciation, thank you. Second, the research has been supported by the Strategic Research Council at the Academy of Finland, for Ethical AI for the Governance of the Society (ETAIROS) project. Third, we would further like to thank all the authors of this book for their significant contributions to understanding emotional experience in relation to technology design, their institutions and the funders who supported them in their work. This book's efforts have additionally been undertaken as a part of a Postdocs in Companies (DIMECC funded PoDoCo) project, by Rebekah Rousi in her time at Gofore. Many thanks for the hours of careful proofreading work undertaken by Kelley Friel. Furthermore, the Faculty of Information Technology and the Cognitive Science (Human Technology Interaction) team at the University of Jyväskylä, Finland, should be thanked for their endless support and embracement of innovative research in cognition, emotions, AI and ethics. Authors of Chap. 5, 'An Innovative Humour Design Concept for Depression', would like to acknowledge the National Social Science Foundation of China for their support of the article's research (Grant No. 14ZDBD155). Acknowledgements are also made by the authors of Chap. 8, 'The role of cuteness aesthetics in interaction', whereby the Fuji-chan device was created as part of a research project supported by a Grant-in-Aid for Scientific Research (KAKENHI project number 15F15052) from the Japan Society for the Promotion of Science (JSPS). The study on the lootbox and croupier design was partly funded by the FWO Travel Grant (2018) and the Short Term Scientific Mission of the COST Action IS1410, call number 5. Finally, the author of Chap. 6, 'The Good, the Bad and the Ugly Graffiti' would like to thank her supervisors Pertti Saariluoma, Professor of Cognitive Science at the University of Jyväskylä; Annika Waenerberg, Professor of Art History at the University of Jyväskylä; and Johanna Silvennoinen, Postdoctoral Researcher, for support and sources of information. This research was supported by research grants from the Finnish Cultural Foundation and the University of Jyväskylä's Faculty of Information Technology.

Contents

Part IV   Ethics and Culture

Editors and Contributors

About the Editors

Rebekah Rousi is a Postdoctoral Researcher of Cognitive Science who has been studying human experience in human technology interaction since 2009. Her research is inspired by her training and practice as a contemporary performance and print-media artist. She is interested in semiotics and language in experience, human-robot interaction and the influences of culture on cognition and emotions. Rousi is currently undertaking a DIMECC funded Postdocs in Companies (PoDoCo) project at Gofore, Finland. Email: rebekah.rousi@jyu.fi

Jaana Leikas is Principal Scientist at VTT Technical Research Centre of Finland Ltd. She holds an appointment as Adjunct Professor in Cognitive Science at Jyväskylä University. Her focus areas and strategic themes include design thinking, social evolution of technology, responsible research and innovation, and technology ethics. In this context, she is particularly interested in how new technologies can best serve the quality of life, changing life goals and lifestyle preferences into advanced age. Email: jaana.leikas@vtt.fi

Pertti Saariluoma is a Professor of Cognitive Science at the University of Jyväskylä, Finland. Saariluoma's work focuses on understanding the psychology behind how people interact with technologies. Saariluoma uses philosophical thinking and argumentation to study processes behind innovation, design, design thinking, user experience and other mental phenomena associated with the cognitive-affective processes of human-technology interaction (i.e., trust in autonomous technology, artificial intelligence etc.) in regards to past, present and future innovations. Saariluoma holds a Ph.D. in Psychology, with expertise in expert thinking, economic and engineering thought errors. Email: pertti.saariluoma@jyu.fi

Contributors

Jose Cañas Bajo is a Postdoctoral Researcher at the Virtual Cinema Lab, Aalto University, Finland. Jose holds a Licentiate in Communication, degree (Vocational school) in Image, postgraduates in screenwriting and a Ph.D. in Cognitive Science. He has also work as a Camera technician, Cameraman, video editor and Scriptwriter. Email: jose.canasbajo@aalto.fi

Gilbert Cockton is an internationally renowned researcher with a career spanning over two decades. During that time he has received almost 220 invitations to present in 22 different countries, published over 220 papers, chapters, books, articles and edited proceedings (selection available from academia.edu), with almost 2500 citations (Google Scholar). He has a broad multidisciplinary background, with an MA/PGCE in History and Human Sciences (Education) and a Ph.D. in Computer Science. His research spans from the theoretical foundations of design and evaluation approaches, to applied work with industry on usability, user experience, accessibility and applications of value-focused design and evaluation approaches. Email: gilbert.cockton@sunderland.ac.uk

Paul Haimes is a researcher in interactive media whose recent focus is on cultural phenomena and aesthetics within the context of digital interaction. He is currently an Associate Professor of Design Science at Ritsumeikan University, Japan. Email: paulhaimes@gmail.com

Jussi P.P. Jokinen is a postdoc at User Interfaces group, Aalto University, Finland. Heworks with computational cognitive models of interaction, including human vision, pointing, memory and emotion. Email: jussi.jokinen@aalto.fi

Jaana Leikas is Principal Scientist at VTT Technical Research Centre of Finland Ltd. She holds an appointment as Adjunct Professor (Docent) in Cognitive Science at Jyväskylä University. Her focus areas and strategic themes include design thinking, social evolution of technology, responsible research and innovation and technology ethics. In this context, she is particularly interested in how new technologies can best serve the quality of life, changing life goals and lifestyle preferences into advanced age. Email: jaana.leikas@vtt.fi

Xueyan Li holds a Ph.D. in Cognitive Science. Her thesis titled Haha Moments—Applying Brain Research to Technology Design explores the development of a technological concept intended to ease the effects of depression. Li combines design science with linguistics, psychology and artificial intelligence. Li is currently a lecturer at the Changchun University of Technology, China. She holds a Master's degree in Educational Science from the Macquarie University, Australia. Email: lixueyan2016@163.com

Stuart Medley is Associate Professor of Design at Edith Cowan University. His research interests include communicating with pictures and comics. He is the author of the book The Picture in Design. Stuart has worked as an illustrator for 20 years, including using comics in design thinking. Clients have included the Imperial War Museums in the UK and Showtime in the US. Stuart is a director of the Perth Comic Arts Festival. Email: s.medley@ecu.edu.au

Mari Myllylä is a doctoral student in Cognitive Science at the University of Jyväskylä. Her research focus is on art experience, cultural convergence and embodied mind in graffiti and murals. Email: mari.t.myllyla@student.jyu.fi

Rebekah Rousi is a Postdoctoral Researcher of Cognitive Science who has been studying human experience in human technology interaction since 2009. Her research is inspired by her training and practice as a contemporary performance and print-media artist. She is interested in semiotics and language in experience, human-robot interaction and the influences of culture on cognition and emotions. Rousi is currently undertaking a DIMECC funded Postdocs in Companies (PoDoCo) project at Gofore, Finland. Email: rebekah.rousi@jyu.fi

Pertti Saariluoma is a Professor of Cognitive Science at the University of Jyväskylä, Finland. Saariluoma's work focuses on understanding the psychology behind how people interact with technologies. Saariluoma uses philosophical thinking and argumentation to study processes behind innovation, design, design thinking, user experience and other mental phenomena associated with the cognitive-affective processes of human-technology interaction (i.e. trust in autonomous technology, artificial intelligence, etc.) in regard to past, present and future innovations. Saariluoma holds a Ph.D. in Psychology, with expertise in expert thinking, economic and engineering thought errors. Email: pertti.saariluoma@jyu.fi

Johanna Silvennoinen has a Ph.D. in Cognitive Science and works as a postdoctoral researcher in University of Jyväskylä, Finland. Her research focuses on cognitive and affective processes in multisensory experiences and on designing for targeted experience goals. Email: johanna.silvennoinen@jyu.fi

Dr. Huili Wang is a professor and Ph.D. supervisor at the Institute for Language and Cognition in the School of Foreign Languages, Dalian University of Technology. Her research interests include psycholinguistics, cognitive neurolinguistics, language and mind. Email: huiliw@dlut.edu.cn

Bieke Zaman is an associate **professor in Human-Computer Interaction/Digital Humanities** and research group leader of the **Meaningful Interactions Lab (Mintlab)**, part of the Institute of Media Studies at the Faculty of Social Sciences, KU Leuven, Belgium and affiliated with imec. Bieke is a regular speaker at both national and international conferences and events. She is member of several academic editorial boards (e.g. Associate Editor International Journal of Child-Computer Interaction), and member of international conference committees such as Interaction Design and Children and CHI Play (Fun and Games). Email: bieke.zaman@kuleuven.be

Chapter 1
Introduction—Feelings Matter

Rebekah Rousi, Jaana Leikas, and Pertti Saariluoma

Abstract Emotions are a hot topic in design, human–computer interaction and any area of business these days. Their significance in areas in which people make choices, decisions and engage in action has been undeniable for at least the last 40 years of psychology and consumer scholarship. What once was an extremely contested, fuzzy and (almost) easily scientifically avoidable area, is now at the centre of everyone's interest. In an era of cognitive computing, artificial intelligence (so-called learning and thinking machines), and optimization, all attention is placed on what makes us human, and the ways in which human thought actually operates. This emotional logic, intentionality and consciousness itself, drive not simply the ways in which individuals process (cognitise) information, but also ways in which society and the built world are structured. Emotions play as much a role in shaping technology design, as they do in the way we experience it. This introduction presents a book that takes many angles towards concretely understanding what it is in design that makes people emotionally experience it in the ways that they do. It introduces the main themes and concepts of the book that include ethics, culture, measurement and design methods. It additionally demonstrates a broader understanding of technology in chapters that investigate graffiti, urban and art experience, filmic experience, architecture and cultural movements. It is hoped that combining this broader cultural-emotional insight into one package will enable readers to connect their design practice and research to the broader system of emotions, culture, ethics, lived experience and technology.

R. Rousi (✉)
University of Jyväskylä and Gofore, Jyväskylä, Finland
e-mail: rebekah.rousi@jyu.fi

J. Leikas
VTT, Espoo, Finland
e-mail: jaana.leikas@vtt.fi

P. Saariluoma
University of Jyväskylä, Jyväskylä, Finland
e-mail: pertti.saariluoma@jyu.fi

© Springer Nature Switzerland AG 2020
R. Rousi et al. (eds.), *Emotions in Technology Design: From Experience to Ethics*, Human–Computer Interaction Series,
https://doi.org/10.1007/978-3-030-53483-7_1

1

1.1 Introduction

Lucretius (96–55 BC: De Rerum Natura) once stated that 'Nothing can be created out of nothing'. Yet in today's experiential market, the sale of intangibles such as emotion constitutes big business. While it might not be obvious to attribute emotions to physical objects such as stone and steel, even these have emotional value. In the past, stone axes have often had decorative features, which can be explained purely on the basis of their emotional value. These days, brands with a high emotional value also often have high monetary value. It, therefore, makes sense to consider the role of emotions from a range of perspectives when designing technology.

People experience their environment and its events through emotional logic. The most ready and commonly cited example at the moment is the COVID-19 crisis. In relation to COVID-19, it may be understood that emotions operate on individual, communal, societal and global levels. Daily news represents the split conflict of the situation, between aiming to decrease the spread and impact of the virus through encouraging, if not forcing, people to stay at home, or addressing the astronomical economic blow the global market is experiencing through enforced consumer behavioural and business operation changes. COVID-19—its name and its image—have become fixtures of our media landscape. When viewing this term and its image, there are mixed feelings about both the phenomenon as well as how it has heavily plagued our daily aesthetics. Yet, at this point in time, if they are not included in some way within current discourse, academic, social, business or design alike, questions begin to arise regarding why it is being ignored: whether or not the material presented is current, and or whether or not the owner of the discourse is attempting to cover over what is currently taking place. The emotional impact generated by such a crisis, therefore, can be interpreted as primal (the global public is certainly in fear for personal health and well-being), social (the impact on social interactions and even the social understanding of the crisis is remarkable) and cultural (it has infiltrated every aspect of our cultural communication and representation, whether symbolically, discursively, economically or politically).

Referring to emotions themselves, on a basic level, emotions render experiences positive, negative or ambivalent. They also focus our attention, reinforce memory and the prioritisation of mentally bound information, and provide meaning-based frameworks with which to categorise phenomena and associated experiential knowledge. Emotions flavour experience and express what is truly important to a person. Thus, emotions are intrinsically connected to human values and needs. Emotions can thus, also be seen as essential for our understanding of right and wrong, meaning that the moral basis, or ethical framework within specific cultures, is equally as important to understand as the emotional connection people have with phenomena in the world around them. Therefore, emotions turn experience into an ethical scale. This is why emotional aspects are so important for technology designers. In this book, we ask as a group of researchers how emotions are (and should be) linked today to our thoughts in design processes, in the experience of final products, as well as in result of emotional dimensions coming together.

The term 'emotion' and ideas about how emotions influence design processes and design experiences have been at the heart of discussions for at least three decades, particularly in relation to human–technology interaction (HTI). Despite the current hype and general acceptance of the importance of emotions when considering humans in relation to technology, it is often overlooked that emotions have caused problems and disagreement for scholars for many centuries. While theorists such as Baumgarten (1936), Descartes (2005) and Peirce (2009) have mentioned the influence of phenomena such as feelings on thought and logic, the association between emotions and their significance in the processes of cognition and behaviour have been largely overlooked. In fact, acceptance of the systematic scholarship of emotions is relatively new, even in the field of psychology (Rousi 2013; Smith and Lazarus 1990). Due to the seemingly fuzzy nature of emotions—what they are, where they come from, why they exist (if they do), and types of (basic) emotions and their cognitive–affective processes—this area of scientific inquiry has been vastly neglected and even rejected.

Major influencers in the history of psychology such as Skinner (1953) have classified emotions as irrelevant epiphenomena; they argued that human thought existed on a behavioural, mechanical, action–reward type level. One reason for the lack of in-depth scholarship on emotions is the difficulty associated with objectively and systematically measuring emotions (Brave and Nass 2007)—their intensity and quality of experience. In other words, as people make sense of their emotions through their own experiential framework and its attributes (qualia, see, e.g. Jackson 1982), there are no tangible or material indications of what actually happens within this highly subjective and relative world of experience. Yet, in a world of rapid technological development, particularly in relation to cognitive information technology (artificial intelligence and machine learning), the role of emotions in logic and thought is becoming ever more important. Thus, it is not simply the emotions of the human actors involved in the design creation and interactional experience process that concerns scientists, but the role that emotions play in logic and reason itself. Cracking the emotional code will determine whether machines can be developed to *think*.

The significance of emotions in relation to cognition was acknowledged in the 1960s and 1970s (Bolles 1974; Lazarus 1966; Tomkins 1962). Yet there were few efforts to intensively scientifically examine emotions and their nature, particularly in the context of technology development (Rousi 2013). Fundamental work in emotion research has taken place since the 1980s. Researchers in psychology have endeavoured to understand emotions, from their core to their physical expression and the types of emotions people experience. Basic emotions, or primal emotional responses, such as love, hate, surprise, disgust and envy, have attracted considerable attention (Ekman 1999; Izard and Ackerman 2000; Frijda 1988; Plutchik and Kellerman 1980; Scherer and Ekman 1984; Smith and Lazarus 1990). These seminal works on emotion from the field of psychology have gradually been brought into research on design, design experience and human–technology (or computer) interaction, often in discussions of user experience (see e.g. Desmet and Hekkert 2007; Hekkert 2006; Saariluoma and Jokinen 2014).

This book further explains, exemplifies and problematises current understandings of the role of emotions in cognition in relation to technology design. Technology design is a vast area covering objects, services and systems that span from drinking glassware to home entertainment devices, mobile computing devices and transportation vehicles. It is important to understand that not only are emotions present in every factor related to technology, but also that everything created and used by human beings is technology and thus, driven by emotions (Mick and Fournier 1998; Thüring and Mahlke 2007). Moreover, the factors and dimensions that influence and define emotions exist not only at the individual level—i.e. whether an individual is directly negatively or positively affected—but also on the cultural, social and societal levels. Culture plays a major role in framing emotional experience, especially in relation to cultural products (design). It also contributes to people's intrinsic and extrinsic motivations and understandings of ethics—i.e. the moral rules and norms that people live by regarding their existence in the world and in relation to other people (Shweder et al. 1993).

Ethics, which shape our sense of right and wrong, therefore, have a major influence on people's emotional experiences (Freeman 2014; Shweder et al. 1993). It is for this reason, for example, that such a substantial dilemma is posed by the COVID-19 crisis. For, on the one hand, lives can be saved if individuals and groups are isolated and the virus is set to run its course within a limited number of individuals. Yet, on the other hand, millions of jobs are at stake, which leads to other health and social problems (depression and anxiety, as well as access to health care for instance). On a national level, thousands of new applications for unemployment benefits are being made each day due to hospitality, tourism and other industry closures. The ethical boundaries rest between people's lives and their livelihoods, which in turn affects life quality once more. These dimensions inevitably also taint whatever is being produced, promoted and consumed in today's world.

From a theoretical perspective, the fields of applied ethics, which involve practical applications of moral considerations, are varied (religious, environmental, organisation, business, research, etc.). They constitute cultural frameworks for understanding fairness and justice, how our behaviour affects those and the world around us and how the world in turn affects us. Thus, the intertwining of ethical understanding with cultural decision-making embedded in design choices explains how and why our cognitive–emotional systems operate as they do towards designed objects, services and systems. From a different perspective, philosopher David Hume (1998/1751) suggested that ethics and moral assessments are emotional by nature, and that in order for any phenomenon to have moral weighting it must also induce distinct emotional reactions. This book discusses the complexity of the relationship between emotions and technology design. The chapters present research that has focused on the influence of national culture in relation to types of design, emotional design innovations and technology as a cultural product representative of (and influencing emotion on) the societal and personal levels.

Emotions in Technology Design comprises four main parts (see Fig. 1.1): (1) Basic Research; (2) Design; (3) Emotions, Culture and Aesthetics and (4) Ethics and Emotions. The themes and topics within the parts are diverse yet all engage with matters of emotions in technology design.

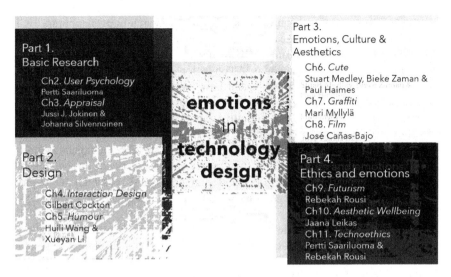

Part 1.
Basic Research
   Ch2. *User Psychology*
   Pertti Saariluoma
   Ch3. *Appraisal*
   Jussi J. Jokinen &
   Johanna Silvennoinen

Part 2.
Design
   Ch4. *Interaction Design*
   Gilbert Cockton
   Ch5. *Humour*
   Huili Wang &
   Xueyan Li

emotions
in
technology
design

Part 3.
Emotions, Culture &
Aesthetics
   Ch6. *Cute*
   Stuart Medley, Bieke Zaman &
   Paul Haimes
   Ch7. *Graffiti*
   Mari Myllylä
   Ch8. *Film*
   José Cañas-Bajo

Part 4.
Ethics and emotions
   Ch9. *Futurism*
   Rebekah Rousi
   Ch10. *Aesthetic Wellbeing*
   Jaana Leikas
   Ch11. *Technoethics*
   Pertti Saariluoma &
   Rebekah Rousi

Fig. 1.1 Structure of *Emotions in Technology Design* according to parts and themes

Part 1 of the book focuses on **Basic Research** approaches to studying and measuring emotions in technology design. These chapters are written from the perspective of Cognitive Scientific research that has applied traditional psychological theories and cognitive theories of emotions to understand the complex relationships between users (people engaging in technology use and consumption), their cognitive–affective processing mechanisms and design. While both chapters within Part 1 focus on basic research, they have differing points of departure. In Chap. 2, on the 'User Psychology of Emotional Interaction' by Pertti Saariluoma, examines and illustrates basic psychological processes involved in human actions and behaviour in relation to technology use and interaction. Saariluoma draws on applied research on ethics, emotions and human-autonomous technology interactions. The chapter provides insight on human values, trust and ethics in relation to design, emphasising the significance of emotions—particularly when people come into contact with new technology. Emotions are highlighted as existing at the core of user psychology while controlling thought processes in technological interaction. Traditions in basic emotion scholarship are operationalised as an emotional-logic platform upon which a considerable amount of human–technology research has been undertaken in recent years. To systematically study these basic emotions, Saariluoma outlines a method of emotional analysis that spotlights the ways in which people respond to technology and its components on a conceptual level. Examples from everyday are used to illustrate how emotional analysis operates in practice.

In terms of a more focused perspective on the study of emotions in user psychology, Chap. 3, 'The Appraisal Theory of Emotion in Human-Computer Interaction', presents Jussi Jokinen and Johanna Silvennoinen's contribution which emphasises the integral role of emotions in all human–computer interaction (HCI).

The chapter discusses the prominence of emotional research in contemporary HCI scholarship, and how the research is instrumental in detecting, predicting and explaining user emotions during HCI processes. The chapter criticises current approaches that lack strong theoretical conviction, without any deeper or robust understanding of what emotions are and how they operate. Jokinen and Silvennoinen's statements are timely and critical as they highlight the importance of detailed understanding between the user's cognitive–emotional system and specific design properties. That is, in order to design *for* emotional experience and to measure its impact, it is (a) necessary to know upon which premise specific emotional reactions will arise; (b) necessary to know how to influence design experience through operationalising these premises within concrete design input and (c) essential to measure emotional design impact through systematically ascertaining and observing these relationships.

Deficits in understanding the theoretical components and explanations of these emotional states will no doubt lead to misunderstandings in interpreting the results and erroneous generalisations. For as long as misconceptions are held in relation to emotions, the ability to solve emotion-related problems is constrained. Thus, Jokinen and Silvennoinen argue that the human cognitive system should be formalised in emotional processes. In order to achieve this, the chapter reviews modern psychological research regarding emotion, particularly the appraisal process. Appraisal refers to the broader cognitive process responsible for evaluating events and phenomena that inform individuals about how to react, behave and cope. This chapter argues that Appraisal theory is particularly instrumental due to its explanatory capacity in relation to HCI design and users' emotional responses. It provides examples of visual user experience and computational modelling of emotions.

Part 2, Design of *Emotions in Technology Design*, presents chapters discussing emotions in the context of applied design. In Chap. 4, 'Research for Designing for Emotions', Gilbert Cockton reflects on the importance of emotions in diverse forms of design, from persuasive design such as advertising and graphic design, to fashion, interior and product design. The chapter analyses the multi-layered nature of design practice and processes to explain the moving and dynamic components of creativity, technique and business that ultimately affect an individual's emotional state. Cockton highlights the heterogeneity of design as practice and praxis, citing the continuum of endeavours from developing complex technological systems to craft-like creativity. Cockton additionally describes interaction design methods and tools intended to assist design teams with designing for emotion.

In addition to these practical means of incorporating emotion into the design progress equation, Cockton is critical towards any 'one-size-fits-all' model. Instead, he argues that due to the dynamic and constantly changing elements from one design case to another, there is always the need to approach interaction design as if it were, to some extent unique. Thus, from the perspective of design professionals, there is truly a need for creativity when designing and developing products based on emotions. Furthermore, truly creative outcomes will always generate unpredictable outcomes across a spectrum of emotions—anticipation for this surprise element on behalf of the design team can be seen as part of the interaction design challenge.

Thus, genuinely novel and creative design work involves a research *for* design that considers work practices, processes and methods in its scope, as well as constructive openness to apply known theory in order to develop new theories. The chapter is critical of approaches to emotions in design that are based on assumptions, as the blind contextual application of generic beliefs on supposed emotions, emotional processing, and subsequent reactions can be detrimental to the design's reach and impact. The chapter additionally stresses a realistic approach to understanding the mismatch between design theory and science, and actual design practice.

In Chap. 5, 'An Innovative Humour Design Concept for Depression', Huili Wang and Xueyan Li present a practical application for emotions in design. They instrumentalise humour as a design element intended to assist individuals suffering from depression. Design Science and Neuroscience are used as disciplinary platforms to inspect alternative concepts for the production of practical technical artefacts to improve the quality of life and well-being of people with depression. Due to the nature of various types of humour and their ability to elicit highly aroused and active positive states—both physiologically and experientially—humour has many potential practical applications for the treatment and prevention of depression. The chapter explains the systematic use of humour to promote physical and mental health, and how it encourages positive social interactions through technology design. Inspection of the neuro-cognitive processing that is activated during emotional responses to humorous design elements provides systematic knowledge with the capacity to inform the creation of intervention tools. This form of study bridges the research gap between Neuroscience and Design Science. It enables the formulation of creative design processes that utilise the positive influences of humour in order to help people regulate and manage their emotional states. Wang and Li propose a humour design concept embodied in the 6Rs (Rethink, Refuse, Reduce, Reuse, Repair and Recycle) model that is applied within the Life-Based Design method. The proposed humour design intervention model is integrated with emerging EEG-based brain–computer interfaces to provide a 'four-in-one' system—real-time recording, instant feedback, effective intervention and constant reinforcement—to lay a more objective and reliable foundation for the implementation of future technical artefacts.

Part 3 examines technology design in the contexts of **Emotions, Culture and Aesthetics**. In Chap. 6, 'The Good, the Bad and the Ugly Graffiti', Mari Myllylä uses graffiti as an example of urban visual communication technology, which spans the spectrum of simple tags to complex masterpieces. Myllylä discusses the specific stylistic format graffiti artists use to convey various meanings. These in turn are intended to incite particular emotional reactions among both viewers and graffiti artists (the producers) themselves. Thus, the chapter focuses on a two-way emotional process that hinges on emotions such as excitement, enjoyment, pleasure, interest, surprise, anger and disgust. Myllylä explains emotional graffiti experience in light of art perception research, elaborating on the underlying cultural dimensions and context that ultimately affect individuals' semiotic readings of the pieces in relation to sociocultural semantic value. Thus, aesthetic emotional responses vary according to multiple dynamic factors that inform the graffiti encounters and interactions. An intersubjective process of reflection and projection between artist and viewer is anal-ysed through the lens of Appraisal theory. Thus, the emotional experience of graffiti draws on factors ranging from the sociocultural and physical environments to the highly personal and subjective experiences, knowledge, and motives of artists and viewers.

In Chap. 7, 'Emotional Film Experience', Jose Canãs Bajo discusses the emotional experiential aspects of technological developments in the film industry. This chapter describes how the evolution of technology such as enriched audio-visual language has affected viewers' emotional engagement. Moreover, through the development of fidelity in the world of film, the idiosyncrasies of one's own lived experience have become more pronounced through the technological medium. Fictional stories come ever closer to reality, which awakens complex cognitive–affective reactions that in turn enhances levels of interest and empathy. Here, Canãs Bajo argues that the role of non-narrative factors such as music and colour have implicit yet significant implications for the influence of emotions in response to the filmic design. Canãs Bajo asserts that ideas and methods from film theory can be effectively applied to studying user experience of HTI in general.

Stuart Medley, Bieke Zaman and Paul Haimes' 'The Role of Cuteness Aesthetics in Interaction', Chap. 8, examines the functional role of aesthetics and emotions in various cultural communicational interaction design settings. With traditions in Japanese design and ideologies that are founded to counterbalance the strict hier-archies and social order of Japan, cute (or Kawaii) is an aesthetic domain that has evolved across the spaces of Japanese culture. Cuteness draws on infantile principles of attractiveness (paedomorphosis) that are said to elicit emotional responses related to nurturing, caring, empathy and compassion (Kringelbach et al. 2016). In Chap. 6, Medley et al. illustrate how cuteness can be utilised to encourage careful behaviour and focus user attention through its emotional rationale. The chapter discusses the outcomes of two projects that focused on cuteness in interaction design. Studies cited in the chapter indicate that context plays a fundamental role in determining the experienced appropriateness of cute design. The empirical studies found that when there is an experienced mismatch of cuteness to context, negative emotions are elicited. This in turn may be detrimental to the intended behavioural response of communicational interaction design. This chapter also highlights the role of *ethics* as an emotional determinant—ethics not simply in designer behaviour, practice and praxis, but also in the aesthetic elements and contextual positioning themselves.

Part 4 concentrates on **Ethics and Emotions** in Technology Design. Chap. 9, 'That Crazy World We Live in—Emotions and Anticipations of the Unexpected in Future Technology Design' examines the cognitive–affective roles of culture and politics and how they interact when anticipating and recognising futuristic design. Rebekah Rousi argues that the radical and the unexpected in future design innova-tions are in fact stylised. Based on these styles, futuristic design elicits specific emotional patterns depending on how the styles and their discourse have been related to specific historical cultural and political developments. Therefore, implicitly Rousi's chapter reflects on the dynamic relationship between cultural and political historical discourse, ethics and emotions. Moreover, the chapter presents an interplay between the cultural and the psychological by illustrating the cause-and-effect relationship that the art movement Futurism had with political violence and fascist rhetoric. Thus, the chapter highlights two components that emotionally influence the experience of future design: (1) the recognisability of future aesthetic style and (2) futurism as a vehicle for alienation and violence.

The first component refers to how future technological design must be aesthetically recognised as such, and thus conform to people's preconceptions of what futuristic design is in form and function in order for it to be accepted as future technological design. The cognitive–affective process implicated in this evaluative exchange is explained through the Most Advanced Yet Acceptable theory (Hekkert et al. 2003). The industrial example of the defunct electronics company General Magic is analysed to demonstrate the role of timing and people's expectations of form and technological capabilities, and how a mismatch of these elements can lead to the collapse of products and even companies. The second component is explained in terms of the ways in which cultural produce (technology design) can be cognitively affectively appraised in relation to its connections with violence, pain and death. That is, while historical developments may be considered to rest in the past, formalistic remnants of these developments remain in the symbolism of the designs. Thus, this chapter importantly discusses the role of culture and cultural discourse in cognition and emotions when considering future technology design.

Jaana Leikas explains the role of aesthetics in emotional design for the elderly in Chap. 10, 'Aesthetic Wellbeing and Ethical Design of Technology'. This chapter uses examples from elderly care homes to illustrate that aesthetics are a central quality of design, and that desired qualities such as the 'feeling of home' are important to elderly people's sense of well-being, both on psychological and ethical bases. Leikas argues that aesthetics are pivotal in evoking emotional states that have the potential to psychologically—and, arguably, physically—help people transition between residences and cope with change and loss, all of which influence their well-being. Focusing on ethics through aesthetics introduces an area of technology design that adopts an ethical approach to addressing valuable design problems such as the meaning and meaningfulness of individual symbolic values in design, as well as the harmonisation and composition of artefacts in the creation of home aesthetics, or 'the feeling of home'. Although this topic is very important given the ageing global population, mainstream HCI and design scholars largely reject it. In order to create a more homely and emotionally healthy aesthetic future, the subject demands thought, discussion and action now.

Finally, Chap. 11, 'Emotions and Technoethics' sums up many of the ideas that have arisen throughout the book. This chapter in particular examines the ways in which technological ethics—or technoethics—can be used as an explanatory vehicle for understanding why people emotionally respond to technology design in the ways they do. The chapter takes on a discursive, cultural and cognitive scientific approach to defining ethics and the Golden Rule—treat others in the way you would yourself want to be treated. Saariluoma and Rousi take up some of the issues mentioned in Chap. 9 to emphasise the symbolic and functional attachment of technology design to original intention and socially generated meaning. Here, examples are drawn from fascist history and current popular technology brands. It also presents ideas on technological ethical neutrality, whereby ethics are attached to intention (consciousness and human thought) rather than being innate features of the designs themselves. Thus, the chapter awakens many controversial issues, and no doubt is set to awaken debate and discussion on a number of the topics it presents. We, the Editors, feel that Chap. 11 opens space for discomfort and problematisation of both framing technological design in relation to ethics and emotions, as well as in regard to understanding the relationship between emotional experience and technology design through the lens of ethics.

The book's goal is to emphasise that technology, cognition and associated emotions are necessary components of human practice and praxis, in which design is an essential action. These practices and associated technologies vary according to the context, intention and human role, yet they are inseparable from (and expressions of) human cognition, emotion and indeed culture. The book presents a variety of cases and perspectives to advance our understanding of emotions in relation to technology design. This is not a guidebook or set formula on 'how to do'. While there are valuable methods, tools and study findings, our intention s to present an object for discussion, debate and to stimulate reactions. For, as Gilbert Cockton (Chap. 4) mentions "[w]here emotions are poorly considered … is typically due to instrumentally focused technologists who want to meet specified requirements to solve a clearly specified problem". The researchers and designers of this book understand that emotions are as organic in nature as they are culturally and contextually situated. A rigid engineer-like or formulaic approach to emotions will never be entirely successful. Yet, isolating, reflecting upon, discussing and arguing the multi-facets of emotions and design properties will aid in knowing not just what emotions may be expected, but why, and by whom.

References

Baumgarten A (1936) Aesthetica. J Laterza and Sons, Bari, Italy

Bolles RC (1974) Cognition and motivation: some historical trends. Cognitive views of human motivation. Academic Press, Cambride, MA, USA, pp 1–20

Brave S, Nass C (2007) Emotion in human-computer interaction. In: Sears A, Jacko J (eds) The human-computer interaction handbook. CRC Press, Boca Raton, FL, pp 103–118

Descartes R (2005) Discourse on method and meditations on first philosophy. Haldane E (trans.), Stilwell: Digireads

Desmet PM, Hekkert P (2007) Framework of product experience. Int J Des 1(1):57–66

Ekman P (1999) Basic emotions. In: Dalgeish T, Power M (eds) Handbook of cognition and emotion. Wiley Ltd, New York, pp 45–59

Freeman D (2014) Art's emotions: ethics, expression and aesthetic experience. Routledge, London

Frijda N (1988) The laws of emotion. Am Psychol 43(5):349–358

Hekkert P (2006) Design aesthetics: principles of pleasure in design. Psychol Sci 48(2):157–172

Hekkert P, Snelders D, Van Wieringen PC (2003) 'Most advanced, yet acceptable': typicality and novelty as joint predictors of aesthetic preference in industrial design. Br J Psychol 94(1):111–124

Hume D (1998 [1751]) An enquiry concerning the principles of morals: a critical edition. In: Beauchamp TL (ed). Clarendon Press, Oxford

Izard CE, Ackerman BP (2000) Motivational, organizational, and regulatory functions of discrete emotions. In: Lewis M, Haviland-Jones J (eds) Handbook of emotions, 2nd edn. Guilford Press, New York, pp 253–264

Jackson F (1982) Epiphenomenal qualia. Philos Q 32(127):127–136

Kringelbach ML, Stark EA, Alexander C et al (2016) On cuteness: unlocking the parental brain and beyond. Trends Cogn Sci 20(7):545–558

Lazarus RS (1966) Psychological stress and the coping process. McGraw-Hill, New York, NY, USA

Mick DG, Fournier S (1998) Paradoxes of technology: consumer cognizance, emotions, and coping strategies. J Consum Res 25(2):123–143

Peirce CS (2009). The writings of Charles S. Peirce, a chronological edition, vol 8. Indiana University, Bloomington, pp 1890–1892

Plutchik R, Kellerman F (eds) (1980). Emotion: theory, research, and experience, vol 1. Theories of emotion. Academic Press, New York

Rousi R (2013) From cute to content—user experience from a cognitive semiotic perspective. Dissertation, University of Jyväskylä, Finland

Saariluoma P, Jokinen JP (2014) Emotional dimensions of user experience: a user psychological analysis. Int J Hum Comput Intraction 30(4):303–320

Scherer P, Ekman P (eds) (1984) Approaches to emotion. Lawrence Erlbaum, Hillsdale, NJ

Shweder RA, Haidt J, Horton R et al (1993) The cultural psychology of the emotions. In: Lewis M, Haviland-Jones M, Barrett LF (eds) Handbook of emotions. Guilford Press, New York and London, pp 409–427

Skinner B (1953) Science and human behavior. The Free Press, New York

Smith C, Lazarus R (1990) Chapter 23—emotion and adaptation. In: Pervin L (ed) Handbook of personality: theory and research. Guilford Press, New York, pp 609–637

Thüring M, Mahlke S (2007) Usability, aesthetics and emotions in human–technology interaction. Int J Psychol 42(4):253–264

Tomkins S (1962) Affect, imagery and consciousness: the positive affects. Springer, New York, NY, USA

Part I
Basic Research

Despite work over the recent few decades devoted to reconciling the emotions-science-technology divide, there has still been a separatist way of viewing emotions across many areas of science and technology. A lot can be owed to the Enlightenment for this Westernised view to treating emotions as fuzzy phenomena. Modern day business and superficial design understanding seem to reinforce understandings of the role of emotions as being the desired result of colour choice and images. The term 'technology' is easily relegated to the realm of information technology, or that, which is steel, mechanical, and/or plastic—devoid of connection to human emotions. Technology can be interpreted as the capacity for action and displaying qualities of capabilities to undertake important tasks. All too often the human component of technology is forgotten. Technology is human. It is created and designed for humans, to serve human intentions and desires, to meet human end goals and states. Thus, feelings and emotions towards, within and surrounding technology are always present. Emotions describe humans' true relationship to technology.

Technology is about what people do. What people do with technology is determined by what they feel is important. The technology they engage in to achieve these actions depends on a variety of factors from what they experience as nice, acceptable, reliable, trustworthy and even economically beneficial (financial value). Feelings and emotions are an inseparable part of the technological world, in human life and the human mind.

The first step towards systematically accounting for and integrating emotional dimensions of technology experience within design involves engaging in robust and valid scientific research. Research in emotions provides opportunities to develop methods and tools that enable the inclusion of emotions into technological thought. This technology-motivated analysis of human emotions is referred to here as 'basic research'. This form of research continuously evolves in relation to current understandings of emotions and how they are applicable to technology and its development. There are sets of emotions and feelings that are more valid for technological development than others. For instance, trust, disappointment, the sense of competence and accomplishment, and perhaps even lust and admiration in specific technological contexts. Other types of emotions including grief, sadness, disgust and fear may not be desired or aspired to in relation to technology design, yet these may be unwanted side-effects when elements relating to technology, its use and associated meaning-making processes (social and cultural discourse) go horribly wrong. The task of basic research and this section is to observe the sub-discourses on emotions, which can have meaningful links to technology use, innovation and development.

Chapter 2
User Psychology of Emotional Interaction—Usability, User Experience and Technology Ethics

Pertti Saariluoma

Abstract Human emotions are decisive in defining what the values of events and phenomena are for a person. This is why emotions are crucial in all our attempts to understand human–technology interaction. Emotions play a key role in all types of design and user-related issues, even in regard to simply whether or not people like a design, and why. Moreover, emotions are additionally important in all attempts to understand ethical issues. Especially, designers should see the explanatory value of emotional states when they pursue to discover why some design solutions function and why others do not. Therefore, it is important to decipher emotions, what they are, and what triggers them, as well as how they can be conceptualized in design research.

2.1 Introduction

All technologies are for people to use to improve the quality of their everyday lives (Saariluoma et al. 2016). Thus, it makes sense to consider how emotions are involved in human–technology interaction (HTI). Emotions define the value of objects for individual people; the quality of life is closely linked to emotional harmony. Therefore, emotions represent a central topic in investigating and designing technical artefacts and technologies.

The study of emotions mainly belongs to psychology; thus, they are psychological phenomena. User psychology is a scientific discourse that investigates the emotions associated with HTI in order to explain (and therefore to understand) why people relate to technologies as they do (Saariluoma and Jokinen 2014). Emotions explain why people do not like to use some types of technology, such as washing machines. For instance, these machines may be ugly, noisy, smelly, unsafe or unreliable. All of these factors generate emotions in users' minds, which in turn explain people's reluctance to use the product.

P. Saariluoma (✉)
Faculty of Information Technology, University of Jyväskylä, Jyväskylä, Finland
e-mail: pertti.saariluoma@jyu.fi

© Springer Nature Switzerland AG 2020

R. Rousi et al. (eds.), *Emotions in Technology Design: From Experience to Ethics*, Human–Computer Interaction Series,
https://doi.org/10.1007/978-3-030-53483-7_2

Technologies may also produce important positive features such as fame, goodwill, brand and positive status. Users feel good about employing technologies that fit with their values, self-image or lifestyle. It is valuable for designers to understand how people emotionally experience their products by studying user psychology, which explicates the emotional experiences associated with a technology or technical artefact.

Psychologically, emotions are phenomena that are very intimately linked to motives. They define why people do what they do in a psychological sense. Therefore, in positive emotional states people pursue certain goals and products, and in negative states they try to avoid them. Thus, emotions can explain many aspects of human motivation.

Liking and *disliking* are not marketing or service design issues; they are modern psychology issues. They are considered important in a number of well-known design paradigms such as Kansei engineering, affective ergonomics, emotional usability and user experience research (Brave and Nass 2009; Hassenzahl 2011; Helander and Khalid 2006; Nagamashi 2011; Norman 2004). The term *dynamic user psychology* refers to the psychology of emotions and motives, which are linked through liking and disliking (Brave and Nass 2009; Hassenzahl 2011; Nagamashi 2011; Norman, 2004; Neisser 1967).

The importance of emotions should not be underestimated. Emotional questions not only concern individual beauty and pleasure; they can affect large numbers of people, and thus the whole society. For example, the Three Mile Island (1979), Chernobyl (1986), Fosmark (2006) and Fukushima (2011) accidents made people feel unsafe; many individuals lost their trust in nuclear power as a result (Friedman 2011; Saariluoma et al. 2016).

Likes and dislikes explain much of why clients accept and adopt some new technologies. The emotional and motivational dimensions of the human mind are central, and they are different from cognition in both their content and their neural representations (Neisser 1967).

Dynamic psychology has traditionally arisen from the clinical personality theories of Freud (1917/2000). However, there is no reason to limit the use of the term to psychoanalysis. In modern psychology, the most relevant paradigm of dynamic psychology is perhaps *positive psychology*. This paradigm analyses issues such as well-being, contentment, satisfaction, happiness and the flow of experience, which have proven to be important in many contexts of *user experience* (Seligman and Csikszentmihalyi 2000). Thus dynamic psychology involves the mental forces explaining why people act as they do.

2.2 Questions of Emotional User Psychology

User psychology introduces typical design questions, such as

- Do certain colours excite people?

- Does the brand of my watch communicate a good image?
- Do people feel brand loyalty?
- Do I feel that it is safe enough to smoke?
- Is the product offensive?
- Do children like their toys?
- Is buying this product moral?
- Is a particular use of a technical artefact ethical?

Emotional interaction research seeks to determine how objects are emotionally experienced, as well as the rational grounds for solving emotional design issues. This process involves investigating general psychological phenomena and using this knowledge to find solutions to technological problems. Following this line of thinking, it is possible to discuss the role of the psychology of dynamic interaction on a scientific level.

A core problem associated with evaluating the user psychology of emotional interactions is analyzing the emotional processes involved in interactions. Designers have to be able to explicate which emotions are relevant—and in what ways—when people interact with technologies. Good analytical practices are necessary to provide clear information about how the design prototypes should be improved.

Saariluoma and Jokinen (2014) presented an example of emotional analysis. They applied a type of psycho-semantic analysis that has been used in Kansei engineering (Nagamashi 2011). Saariluoma and Jokinen (2014) used a Likert-type survey in which subjects' basic emotions were used as dimensions. They employed multidimensional scaling to identify an emotional dichotomy of competence vs. frustration: if people were able to use a technology they felt competent, but if they failed they became frustrated.

2.3 Emotions and the Mind

Emotions form a fundamental mental process in the human mind (Ekman 1999; Frijda 1986; LeDoux 1998; Rolls 2000). They serve as an action control system. Since emotions are always present in human information processing, they also represent a fundamental aspect of HTI. Progress on these issues is essential to the development of the field. User psychology should formulate good methods and concepts of emotional analysis for HTI and reliable practices to apply these methods in the product development process.

Emotions play a valuable role in explaining people's behaviours (Ekman 1999; Frijda 1986, 1988, 2007; Oatley 2006). Evolutionarily, emotions establish a primitive neural system, and therefore the main characteristics of many human emotions can be found in other animals as well (Darwin 1872/1999 Panksepp 1998). Emotional processing is mainly centred in the subcortical areas in the brain, which are typical to evolutionarily early systems (McClean 1990; Rolls 2000). However, emotional processes are fundamental in controlling many behavioural processes. Cognition can explain why something is *worth* fearing but not fear itself.

Emotional systems operate holistically, and feature equally important physiological, psychological and social characteristics. There are numerous ample reviews and overviews of emotional processes, thus, in this book, we do not review basic theories of emotions in detail. If interested in further reading, we recommend that readers refer to following literature (Ekman 1999; Frijda 1986, 1988; Oatley 2006; Panksepp 1998; Power and Dalgleish 1997; Rolls 2000; Lazarus and Lazarus 1994). In any case, human–technology interaction research should examine emotions holistically.

A ready example of a long line of holistic thinking of emotions is hierarchical affect theory (Laros et al. 2005; Watson and Stanton 2017). In this theory, it is essential to divide emotions into groups on the grounds of positivity vs. negativity. Thus, the theory raises the issue of conceptual dimensions of hierarchy which are essential in defining emotions. In this chapter, the analysis of conceptual dimensions will be further discussed and elaborated.

Emotional state—how people feel at a certain moment in time—is a foundational concept in any analysis of the emotional mind. It illustrates the emotional aspect of a prevailing mental representation. Emotional states are thus aspects of mental states. They bring additional knowledge regarding the mental content of representations. These aspects must be explicated one at a time to understand how emotions operate with mental representations. Emotional states have four main basic dimensions.

The first is the *intensity* of emotions, which is called arousal (Kahnemann 1973; Power and Dalgleish 1997). There is a graded structure typical to all human emotions, and the intensity or level of arousal is used to express how emotions vary. Frustration and hate, for example, are emotional states towards an object, person or state of affairs (Norman 2004). Irritation can be seen as a less intense version of the same emotion.

The level of human performance depends on the intensity of the prevailing emotional state (James 1890; Kahnemann 1973). When arousal is very low, the human performance is not intense. When arousal increases, the performance capacity also increases. However, after a certain point, increasing arousal no longer improves the level of performance, which starts to weaken. When frightened, for example, people seldom act rationally because their cognitive capacity has diminished. Understanding this principle, called the Yerkes–Dodson law, and other aspects of arousal support the design of technical artefacts and systems that are meant to be used in critical situations or under pressure (Kahnemann 1973).

The second dimension of an emotional state is the *extent* of an emotion. A person's response to a certain incident may be very short, such as a rapid reaction to a surprise (Power and Dalgleish 1997). In technology, design surprise is sometimes referred to as the *wow effect*. An emotional state can also last for weeks, as in *moods* or attitudes (Power and Dalgleish 1997). A good example of a long-lasting emotional response tendency is *temperament*, which forms in infancy and early childhood.

The third property of emotional states is their content. For instance, joy is a positive emotion and sorrow is its negative counterpart. Consequently, technical artefacts may include different aspects of emotional content, for example, when considering the feelings aroused by the design of a product. In this kind of analysis, two new dimensions of human emotion become relevant: *emotional valence* illustrates the negativity or positivity of the emotion, and *emotional theme* refers to its general content (Lazarus 1991; Saariluoma and Jokinen 2014). Emotions can be positive or negative (Lazarus and Lazarus 1994; Spinoza 1675/1955). Emotional valence is strongly connected to the pleasantness or unpleasantness of the emotional state, and thus valence is essential in defining the desirability of emotional contacts. A positive emotional contact is an important design goal in HTI (Jordan 2000; Hassenzahl 2011). Emotional theme refers to what the emotion is (e.g. joy, disgust). They are activated by different circumstances and mental states. Thus joy refers to a feeling of happiness and disgust to a distasteful or unpleasant feeling. It can be related to smells and foods, but it can also refer to experiencing social habits, for example.

An example may clarify the difference between valence and theme. The feeling of joy associated with using technologies usually embodies positive emotions such as wellness, commitment and positivity. Trust is also a positive emotion that can refer, for example, to human reliance on a given technology or its users. Thus, these two emotions—joy and trust—have the same positive valence but different themes. Hence, the *theme* defines the content of emotional states in a more sophisticated manner than valence.

Emotions determine the subjective meaning of a situation to an individual; they are closely connected to human action (Frijda 1986, 1988). They indicate the personal meaning of the target and the possible actions that people might take (Frijda 1986, 1988; Lazarus and Lazarus 1994). The feeling of fear, for example, makes people flee, whereas curiosity often results in approaching a target. Because these emotional characteristics are built into HTI processes, emotions are a critical element of situational representations.

Emotional responses to things or incidents are not static by nature; they change over time and continue to develop over a person's lifetime (Lazarus and Lazarus 1994; Power and Dalgleish 1997). For instance, people who reacted hastily or aggressively to certain incidents in their youth may behave moderately and calmly in the same situations when they are more mature. This process of emotional development is called *emotional learning.*

Emotional learning processes change the content of emotions stored in the memory—so-called emotional schemas—which people use when they select information to retain during the perception process and build memory representations. Emotional learning is an important aspect of people's relationship to technologies. A user who once regarded mobile services as redundant may later become an advocate of this kind of technology after having learned to use and understand its practical value. In this case, a change in an emotional meaning is explained by a change in the content of emotional schemas that are used to construct emotional states through apperception.

2.4 Cognition and Emotions

Cognition refers to how people process information. It refers to perceiving, attending, languages and thinking. These processes create mental representations and illustrate to people how things are in the world. Of course, how people cognitively represent the prevailing situation around them affects how they emotionally encode a situation. The process that associates emotions with cognition is called *appraisal* (Frijda 1986, 1988, 2007).

If parents see a lion approach their child, they normally feel fear, which is an emotion rather than a cognitive phenomenon. If the lion is in a cage, they may find the situation amusing. Cognitive analysis helps people understand whether to be afraid in a particular situation.

Emotional states are not randomly activated. Their activation is based on an individual's understanding of the prevailing state of affairs. If the situation is cognitively risky, this stimulates danger-related emotions such as excitement, fear and courage. Emotional representations are constructed in the human mind, based on cognitive content (Frijda 2007).

Appraisal is a core process in the *psychology of emotions* (Frijda 1986, 1988, 2007). It forms the emotional dimension of mental representations, which defines the value of a situation to the person experiencing it (Frijda 1986). Emotions associated with technologies must be empirically defined in HTI. People are individuals: one user may feel anger while another may feel guilt in the same situation. Emotional states also depend on the context, but there is no conceptual way to study HTI situations. It is therefore essential to empirically define the emotions associated with interaction use cases.

Linking cognitions and emotions is a central issue in HTI design. If people cannot reach their goals using the technical artefacts they have at hand, they become frustrated (Saariluoma and Jokinen 2014). The overall appearance and colours of a technology product affect the mind, as do pictures and 'gestalts' (Norman 2004). The mere image of the product can be emotionally very important, which is why considerable attention is paid to semiotics and art design. The associations of pleasure and displeasure are vital to a product's success.

In the appraisal process, human cognitive representations are connected to active emotional states. If users do not think they can learn to use an e-learning technology, they become frustrated (Juutinen and Saariluoma 2007). As a consequence, they will have generally poor experiences. Thus, cognitive assessments can generate emotional frustration regarding the whole action.

These dimensions are not always rationally connected, as people may emotionally represent situations in an inadequate manner; in the case of technology usage, they may even misrepresent the technologies or their uses both cognitively and emotionally. This may shape an individual's feeling of self-efficacy—that is, their confidence in their ability to successfully perform a task (Bandura 1997). Mistaken beliefs about one's own incapacity may even lead to a self-fulfilling misconception.

Repeated failures create a negative atmosphere and lower self-efficacy, whereas success in using technology improves self-efficacy and generates positive feelings and pride. This makes people more willing to accept, use and train to use new technologies to achieve their goals (Juutinen and Saariluoma 2007). The example illustrates how appraisal is significant in indicating personal meanings of technologies to people, how people differentiate their emotions, as well as their cognitive and physiological behavioural responses. Thus appraisal connects cognitive evaluation with emotions and actual human actions. Individual preferences and action choices are constructed based on representations made during the appraisal process.

The challenge of appraisal-based HTI research is defining what kinds of cognitions activate certain kinds of emotional states. When people interact with a technology, they create human mental representations and corresponding emotional representations. These cognitive and emotional representations control human actions and define whether people like and accept a technology. Thus even a small emotionally misplaced detail in a user interface can easily decrease the value of the technology in people's minds.

2.5 From Emotions to Ethics

Interacting with technologies not only means that people can—and will—use technical artefacts. It is also important to ask how and why they use them (Saariluoma et al. 2016). Since ethics and emotions are related, it is important to discuss technology ethics in the context of emotional interactions.

Technologies are mostly ethically neutral. While kitchen knives are normally used to prepare food, they have also been used to kill people. Thus the focus when investigating ethical interactions with technologies is how a technology is *used*. It is important to think carefully about the ethical issues associated with using technologies, for instance, who has the right to own military class weapons, or should nuclear and coal power be abandoned. A special problem is artificial intelligence (AI) and autonomous technologies, because they have great performance capacities and have been ethically controlled to some degree.

Technology ethics focuses on how particular technologies can be used for good or bad. The role of emotional research in ethics involves defining what is good and bad. This is an emotional process in the human mind (Ayer 1936; Hume 1738/1968). People experience different situations and classify them as good or bad. The routes to these situations and the tools used to respond to them can thus be classified as either positive or negative. Through social discourse, people create the norms of social life (Habermas 2018).

Thus the ways in which people use technologies, the emotional experiences of consequences and social discourses constitute the norms for how to use technologies. Understanding emotions is a key point in this process: human cognitions tell people how things *are*, but emotions decide how they *should* be.

2.6   Emotions in HTI Design Discourses

There are some common concepts and discourses in analyzing emotions in interactions with technologies (Hassenzahl 2011; Helander and Khalid 2006; Nagamashi 2011; Norman 2004; Jordan 2000). Some of them discuss emotional features in products that activate users' emotional states, while others are interested in what takes place in users' minds and what the main characteristics of technology-relevant emotional states are.

Emotions have always been important to consider when designing technologies. Stone age objects often have features that cannot be explained on the basis of their use and usability only. They have unnecessary decorations, for example. Indeed, Plato paid specific attention in the tenth book of state to the beauty and purposefulness of technical objects such as flutes (Platon 601c–d).

In the twentieth century, a major opening to emotion-based thinking was Kansei engineering (Nagamashi 2011). Other paradigms include emotional and affective usability research (Helander and Khalid 2006; Norman 2004), user experience research (Hassenzahl 2011), affective design, pleasurable design (Jordan 2000) and entertainment design (Rauterberg 2010). These intimately linked scientific paradigms have illustrated the relevance of emotions in technology design (Helander and Khalid 2006).

Emotional interaction design is a generally recognized design discourse (Brave and Nass 2009; Hasenzahl 2011; Norman 2004, Saariluoma et al. 2016). Considerable work has been invested in finding ways to design products with excellent emotional interaction properties (Nagamashi 2011). A good explanation of the importance of emotional products is its ability to conquer consumer markets.

Brands, for example, are used to create particular emotional states (Stenros 2005). An example of a paradigm that renovates traditional art design thinking is Scandinavian design. Its minimalist, functional and practical forms are seen to reflect people's relationship with the wild. These forms can be found in the works of designers such as Arne Jacobsen, Poul Henningsen and Alvar Aalto. The pursuit of the ideals of functionality and practicality can also be seen in the designs of information technology.

Emotional interaction does not concern only the immediate use of a technology. Technologies should also respond to emotions that are relevant for human life, such as trust and confidence. People should be able to trust that a given technology operates as expected, and that it will not create or exacerbate any practical or ethical problems. For example, when AI systems fail to be safe, this will decrease audience trust in technology, which may slow the development speed of the whole field. Negative trust in a particular technology can easily lead to technophobia (Brosnan 2002). For example, poor usability may cause stress. Technophobia can disappear as a consequence of direct or indirect positive emotional experiences.

Emotions are present all the time. Even neutral states have emotional values, and their influence is widespread in HTI. Much of human behaviour must be explained in emotional terms. Techno-dependencies provide an extreme example of emotion-related behaviour. Many people become game dependent as a result of playing games or surfing the net. Internet dependency is classified as a psychiatric illness that requires professional intervention. In South Korea, it has been assessed as one of the most serious public health issues, as it has been claimed to be connected to, for example, childhood obesity.

Knowledge of the nature of people's relevant emotional states can be used to generate ethically responsive design goals for technologies, such as personal health monitoring systems. The field of emotional user psychology opens up an extensive set of design questions and challenges that must be seriously investigated.

2.7 Explanatory Emotional Design

The main goal of user psychology is to analyze, explain and design human interaction phenomena in concepts of psychological knowledge and methodology. Emotional user psychology investigates interaction phenomena, which can be analyzed and explained on the grounds of the psychological knowledge we have about human emotions. Of course, through the appraisal process relevant cognitive analyses can also be applied to solve users' psychological problems.

Explaining and analyzing are the ideal functions of user psychological knowledge in technological design. Analysis enables designers to conceptualize the problem correctly, while explaining helps them answer 'why' questions. For instance, why an aircraft blew up or a bridge collapsed are typical questions in technical investigations.

Explaining is thus a common activity in technical design thinking. For example, there is a simple technical explanation based on natural laws why a car radiator breaks in sub-zero temperatures: the water in the radiator turns to ice, and the force of this expansion is too strong for the radiator to withstand (Hempel 1965). This problem can be fixed by using glycol to change the freezing point. Thus the explanation is linked to the problem, and solving it directs the designer towards a concrete solution. Explaining, converging and predicting form a chain of design thought.

There is a fundamental difference between traditional engineering and current HTI design practices. HTI design processes are generally dominated by intuitive procedures such as the free and creative generation of ideas, visioning and user testing. In these, explanatory practices typical to natural science have only a minor role or are absent, although they have the potential to place HTI design on a more solid foundation (Saariluoma 2005; Saariluoma and Jokinen 2014; Saariluoma and Oulasvirta 2010).

Explanations of user psychology should be based on psychology. The main challenge is to unify design problems, psychological methods and theories, and explanatory grounds in a sense-making manner.

Figure 2.1 shows that a designer binds the interaction problem and relevant scientific information together to generate a solution. Each design thought sequence of this

Fig. 2.1 General explanatory framework

type can be seen as a separate explanatory framework, but scientifically grounded design processes are generally characterized by this schema.

The core concept of the explanatory process is coherence. This means that the *explanandum* (the phenomenon to be explained) and the *explanans* (the explaining phenomenon) should be coherent. For example, the superiority of good brands over unknown ones can be analyzed in terms of trust and evidence of past experiences. The structure of the explanatory argumentation is similar to Hempel's radiator illustration.

Even in ethical design and responsible research, innovation processes and emotions are vital. Good and bad feelings help people decide whether a situation is worth pursuing. Ethical rules are derived from such analysis via social discourse, which helps explain social and legal norms. The denial of drugs, a kind of technology, is based on their unpleasant long-term consequences. The emotional states associated with these consequences explain why the related laws are so restrictive.

Psychology and design both have theoretical and practical features. In current design thinking, practicality can only be found by unifying science and design. *Unification* is the key to understanding the function of scientific knowledge in design thinking. It is well known that science and design have multiple connections, and that they are at their best when they are combined. However, precisely defining the connections between the two ways of thinking requires extending the circle of core issues.

Emotions are an important aspect in studies of the human mind; they also have a role in HTI design (Power and Dalgleish 1997; Norman 2004; Saariluoma et al. 2016). Psychologists and sociologists have made considerable progress in studying human emotions (Frijda 1986; Power and Dalgleish 1997). They are interested, for example, in the kinds of reactions that emotional words might inspire in people. Designers, for their part, incorporate emotional features into their systems to make them more appealing and more sellable (Norman 2004).

This chapter argues that emotions should form one of the main grounds in designing technologies. There are phenomena that can be investigated, analyzed and explained based on current knowledge about human emotions. However, when researchers and designers appreciate the importance of basic human research concepts in understanding HTI, it will be possible to conscientiously focus on these themes and make progress.

References

Ayer A (1936) Language, truth, and logic. Gollancz, London

Bandura A (1997) Self-efficacy: the exercise of self-control. Freeman, New York

Brave S, Nass C (2009) Emotion in HCI. In: Sears A, Jacko JA (eds) Human-computer interaction: fundamentals. CRC Press, Boca Raton, FL, pp 53–68

Brosnan MJ (2002) Technophobia: the psychological impact of information technology. Routledge, London

Darwin C (1872/1999). The expression of the emotions in man and animal. Fontana Press, London

Ekman P (1999) Basic emotions. In: Dalgleish T, Power M (eds) Handbook of cognition and emotion. Wiley, Chichester

Freud S (1917/2000) Vorlesungen zur Einführung in die Psychoanalyse. Fischer, Frankfurth am Main

Friedman SM (2011) Three mile Island, Chernobyl, and Fukushima: an analysis of traditional and new media coverage of nuclear accidents and radiation. Bulletin of the Atomic Scientists 67:55–65

Frijda NH (1986) The emotions. Cambridge University Press, Cambridge

Frijda NH (1988) The laws of emotion. Am Psychol 43:349–358

Frijda NH (2007) The laws of emotion. Erlbaum, Mahwah, NJ

Habermas J (2018) Diskursethik (Discourse ethics). Surkamp, Frankfurt am Main

Hassenzahl M (2011) Experience design. Morgan & Claypool, San Rafael, CA

Hempel C (1965) Aspects of scientific explanation. Free Press, New York

Helander M, Khalid HM (2006) Affective and pleasurable design. In: Salvendy G (ed) Handbook of human factors and ergonomics. Wiley, Hoboken, NJ, pp 543–572

Hume D (1738/1968) A tretise on human nature. Dent, London

James W (1890) The principles of psychology. Dover, New York

Jordan PW (2000) Designing pleasurable products: an introduction to the new human factors. CRC Press, Boca Raton, FL

Juutinen S, Saariluoma P (2007) Usability and emotional obstacles in adopting e-learning: a case study. Paper presented at the IRMA International Conference, Vancouver, Canada

Kahnemann D (1973) Attention and effort. Prentice-Hall, Englewood Cliffs, NJ

Lazarus RS (1991) Progress on a cognitive-motivational-relational theory of emotion. Am Psychol 46:819–834

LeDoux J (1998) The emotional brain: the mysterious underpinnings of emotional life. Simon and Schuster, New York

Lazarus RS, Lazarus BN (1994) Passion and reason: making sense of our emotions. Oxford University Press, Oxford

Laros JM, Benedict J, Steenkamp (2005) Emotions in consumer behavior: a hierarchical approach. J Bus Res 58:1437–1445

MacLean P (1990) Triune brain in evolution. Plenum Press, New York

Nagamashi M (2011) Kansei/affective engineering and history of Kansei/ affective engineering in the world. In: Nagamashi M (ed) Kansei/affective engineering. CRC Press, Boca Raton, FL, pp 1–30

Neisser U (1967) Cognitive psychology. Appleton-Century-Crofts, New York

Norman D (2004) Emotional design: why we love (or hate) everyday things. Basic Books, New York

Oatley K, Keltner D, Jenkins JM (2006) Understanding Emotions. Blackwell, Malden, MA

Panksepp J (1998) Affective Neuroscience: the Foundations of Human and Animal Emotions. Oxford University Press, Oxford

Power M, Dalgleish T (1997) Cognition and emotion: from order to disorder. Psychology Press, Hove

Rauterberg M (2010) Emotions: the voice of the unconscious. Entertainment Computing-ICEC 2010. Springer, Berlin, pp 205–215

Rolls ET (2000) Precis of the brain and emotion. Behav Brain Sci 23:177–191

Saariluoma P (2005) Explanatory frameworks for interaction design. In: Pirhonen A, Isomäki H, Roast C, Saariluoma P (eds) Future interaction design. Springer, London, pp 67–83

Saariluoma P, Cañas J, Leikas J (2016) Designing for life. Macmillan, London

Saariluoma P, Jokinen JP (2014) Emotional dimensions of user experience: a user psychological analysis. Int J Hum Comput Interaction 30:303–320

Saariluoma P, Oulasvirta A (2010) User psychology: re-assessing the boundaries of a discipline. Psychology 1:317–328

Seligman MEP, Csikszentmihalyi M (2000) Positive psychology—an introduction. Am Psychol 55:5–14

Spinoza B (1675/1955). Ethics. Dover, New York

Stenros A (2005) Design revolution. Gummerus, Jyväskylä

Watson D, Stanton K (2017) Emotion blends and mixed emotions in the hierarchical structure of affect. Emot Rev 9:99–104

Chapter 3
The Appraisal Theory of Emotion in Human–Computer Interaction

Jussi P. P. Jokinen and Johanna Silvennoinen

Abstract This chapter reviews the appraisal theory of emotion and how it has been employed in human–computer interaction (HCI) research. This theory views emotion as a process that evaluates the subjective significance of an event. We demonstrate the usefulness of the perspective for HCI, as emotion is defined in terms of the events of the task environment and the goals and knowledge of the subject. Importantly, the appraisal theory ties these factors together in a cognitive appraisal process order to explain the variety of subjective emotional experiences. This is important for two reasons. First, a strong theoretical commitment allows researchers and designers to derive testable hypotheses from the theory. Second, only a theory that ties together goals, knowledge and emotion can explain the behaviour and experiences of users, who often have multiple—and at times conflicting—goals and motivations that may dynamically change in response to events in the environment.

3.1 Introduction

Emotion is present in virtually all uses of technology. It is, therefore, not surprising that the study of users' emotions is a large and ever-growing subfield of human–computer interaction (HCI) research. This research has proposed multiple theoretical and empirical approaches to detecting, predicting and explaining users' emotions during interactive tasks. However, these approaches are often agnostic to any strong theoretical commitments about what emotion *is*. Yet, if theoretical assumptions are not explicated in operationalisations of user emotion, this will confound the results

J. P. P. Jokinen
Department of Communications and Networking, Aalto University, Espoo, Finland

J. P. P. Jokinen (✉) · J. Silvennoinen
Faculty of Information Technology, University of Jyväskylä, Jyväskylä, Finland
e-mail: jussi.p.jokinen@helsinki.fi

J. Silvennoinen
e-mail: johanna.silvennoinen@jyu.fi

© Springer Nature Switzerland AG 2020
R. Rousi et al. (eds.), *Emotions in Technology Design: From Experience to Ethics*, Human–Computer Interaction Series,
https://doi.org/10.1007/978-3-030-53483-7_3

and make argumentation imprecise. This limits the capacity of HCI research to solve emotion-related problems.

We argue that formalising the role of the human cognitive system with the emotion process can help overcome this limitation. To this end, we review the contemporary psychological research on emotion, with an emphasis on the appraisal process. The *appraisal theory* defines emotion as a partially cognitive process that evaluates the significance of an event and an individual's ability to cope with it. We discuss the use of theories related to emotion in HCI and demonstrate how the appraisal on account of emotion can improve our ability to understand users' emotions. We explore the implications for different types of studies of user emotion using examples from experimental research on visual and emotional user experience.

When users approach interactive technologies, it is necessary to understand their goals and knowledge in order to explain their behaviour. For instance, an individual may approach a ticket vending machine with the intention of buying a ticket. If the user has accomplished this goal previously, she understands the interactive steps that must be accomplished, and can, therefore, finish the task faster than a novice user with no knowledge of how the interface works. The emotions that these users have during the interaction are conditioned on their goals and knowledge. For example, if the user encounters a problem while buying the ticket, her emotional response is conditioned on her ability to overcome the problem. A more experienced user might know the solution to the problem, but feel frustrated that a bad interface design makes her task more cumbersome. Conversely, a novice user might feel distressed because she does not know how to overcome the problem, and may fail to buy the ticket (perhaps for a train that is leaving soon). In these examples, it is clear that explaining emotion requires referring to the users' goals and knowledge, and to the event of the interaction. The theoretical task of analysing the psychology of emotion is to explain how these factors influence the users' emotional responses.

We argue that an appropriate theoretical approach to analysing the psychology of emotion involves using the appraisal theory of emotion, which posits that emotion is a continuous appraisal process that evaluates the subjective significance of an event. The key in this evaluation is to recognise that emotions are responses to how events are evaluated within the subject's meaning structures (Arnold 1960; Frijda 1988; Lazarus 1966; Scherer 2009). Analysing these meaning structures will help explain an individual's emotional responses to a given event. A person's appraisal of something as 'good' and approachable or desirable, instead of 'bad' and avoidable or undesirable, is a function of their conceptions of good and bad. This in turn depends on what they know about the situation and its circumstances, as well as their personal preferences and goals.

The appraisal theory analyses emotion as a partially cognitive phenomenon, in which information from multiple sources is processed during the appraisal (Smith and Kirby 2001). It also integrates multiple components of emotion, such as physiological responses, motivation, categorisation and labelling (Scherer 2009). This is important for employing the theory in HCI research, which has a long methodological tradition of collecting users' subjective and physiological responses to the use of technologies (Jokinen 2015b). In this context, appraisal theory has the potential to integrate the currently theoretically disparate strands of emotion research in HCI. Moreover, because the theory presents emotion as a causal system from an initial stimulus to a corresponding emotion and its behavioural effects, it provides a way to assess counterfactual scenarios. The ability to predict human responses in different types of 'what if' scenarios is paramount for any interface and task design that claims to be based on scientific thinking.

3.2 Background

Although emotion has always been the source of philosophical debate, it has only recently been examined scientifically. Even after the birth of modern psychology as the science of the human mental life in the nineteenth century, research was mostly focused on human cognition; emotion received less detailed and thorough analysis (Baddeley 2007). The theoretical foundations of all current major psychological theories of emotion were laid in the second half of the twentieth century (Arnold 1960; Ekman and Friesen 1971; Russell 1980), but detailed models and rich empirical data were not developed until decades later (Izard 2007; Russell 2009; Scherer 2009; Jokinen 2015a). Although we focus on the appraisal theory of emotion, we briefly describe other major theories for comparative purposes.

3.2.1 Basic Emotions Theory

To study the hypothesis that emotion is based on evolutionary development and is, therefore, universal in humans, Ekman and Friesen (1971) researched two Oceanic Neolithic cultures. Both cultures were isolated from any Western influence, and were, therefore, good subjects for the study of the universality of emotion. The researchers told the study participants stories that involved emotions (happiness, anger, sadness, disgust, surprise and fear), and asked them to identify which picture of a facial expression best fits with each story. The participants largely identified the intended emotions, which supported the universality of emotion hypothesis.

Theories that state that emotion has a psychobiologically universal pattern are called discrete or basic emotion theories (Ekman 1992). These theories maintain that each set of discrete emotions has a distinct pattern of bodily change, a physiological response and antecedent events. By emphasising the word 'basic', the proponents of these theories claim that (1) there is a fixed number of 'basic emotions' that differ from each other and (2) these emotions are 'basic' because each of them has adaptive value for human life (Ekman 1999). Researchers have different conceptions of which emotions are 'basic', but they generally agree that there are relatively few such emotions.

Discrete emotion theories are often used in the subfield of HCI called affective computing, in which researchers seek to establish a connection between a user's physiological responses and emotional states (Picard 1997). Experimental research in affective computing often reports correlations between self-reported or inferred emotions and physiological measurements, such as heart rate (Drachen et al. 2010), galvanic skin response (Mandryk et al. 2006) or pupil size (Partala and Surakka 2003). However, there are still no universal solutions for the original challenge posed for affective computing—developing an automated emotion-detection machine. Indeed, meta-analyses of the psychophysiology of emotion have yielded mixed results with regard to our ability to detect discrete emotions using neuroscientific instrumentation (Barrett 2006; Cacioppo et al. 1997, 2000; Shiota et al. 2017).

3.2.2   Core Affect Theory

Russell (1980) was interested in finding latent structures in the ways that people categorise emotions, and asked participants to group emotions by similarity. By analysing the emerging latent structures, Russell demonstrated how people often represent emotions using two dimensions, valence (pleasantness) and arousal (activation). This can be called the circumplex model of affect; it places each emotion in a circle according to its valence and arousal. For instance, sadness and pleasure are different due to valence, and sleepiness and excitement are different more due to arousal (although in both cases, there is also a noticeable difference in the other dimension).

Using the circumplex model of emotion, Russell and Barrett (1999) studied how the relationship between the physiological emotion process and affective experience. They proposed that the human emotional system can be considered a fundamental element of emotion, which they called core affect. Core affect is the simplest part of the emotion process that is accessible to the human consciousness (Russell and Barrett 1999). It combines the component values of valence and arousal, and can be identified as a single point in the circumplex—but not one of the emotion words that can be used to describe that emotion and located on the circumplex, as it is an elementary concept. Therefore being in love, for example, can be located on the pleasant side ("feeling good") of the horizontal axis of the circumplex, but love is not a core affect itself. Part of being in love is feeling good and at least somewhat aroused, and this is the core affect; however, love also entails other components (for example, it is usually directed at someone).

The core affect approach to emotion is very popular in HCI, because it provides a framework for classifying emotions that is easy to use and understand. One useful and often applied method is the self-assessment manikin (SAM) scale (Bradley and Lang 1994), which lets participants rate their emotional state using pictorial scales for valance and arousal (e.g. Zimmermann et al. 2006). Thanks to the fairly simple two-dimensional physiologically interpretable construct, core affect is also often used in affective computing. Affective computing research has identified physiological correlates for both arousal and valence (Lichtenstein et al. 2008), although as noted above, these correlations are often not robust across task domains and experiments (Cacioppo et al. 1997, 2000).

3.2.3   Appraisal Theory

The multiple theories about the emotional appraisal process all assume that emotion is related to the evaluation of an event (in the environment or within the individual), which implies it is a complex organised system (Smith and Kirby 2001; Scherer 2009). Further, appraisal is assumed to relate to the desirability or avertability of the appraised stimulus, which assigns an adaptive function to emotion (Arnold 1960). Emotions are elicited in situations that have adaptive significance to the individual; emotions prepare and motivate the individual, and help acquire and filter relevant information (Scherer 2009).

The emotion process can be categorised into four distinct, causally related, components: motivational changes, physiological response patterns, a central representation and verbalisation (Scherer 2009). It might be tempting to conceptualise emotion as the outcome of the appraisal process, but it is better to define it as the appraisal process itself, since it is present in all of the four components. For example, fear might be associated with a motivation to flee and an increase in heart rate, as well as with a conscious experience of being afraid. These components interact with each other, for example, the experience of fear may not only increase a subject's heart rate; the subject may perceive this increase, which changes the experience of the situation and may activate some coping mechanisms—and subsequently alter his or her physiological response. This means that emotion is a multilevel process, not a state. The confusion between these two arises from conflating subjective feeling (which is a single part of the emotion process) with the entire emotion process.

The appraisal theory is not necessarily in conflict with the theories of basic emotion and core affect described above. Rather, these can be subsumed into appraisal theory. The physiological component of appraisal theory can account for the propositions of basic emotion theories, and the core affect theory describes one way in which emotions can be represented, categorised and verbalised. Methodologically, one can operate within the assumptions of appraisal theory and still conduct research on emotions in HCI using psychophysiological methods, such as in affective computing, or collecting subjective emotion data using the SAM scale (Jokinen 2015b). Importantly, appraisal theory also allows the researcher to make hypotheses about the causal evaluative processes of the observed results. The next section gives examples of how this can be accomplished.

3.3   The Appraisal Theory in HCI: Case Examples

3.3.1   Visual Experience

The scope of HCI research has broadened in recent years. The focus is no longer merely on sensemaking processes including research on functionality and usability. An increasing amount of research is seeking a holistic understanding of HCI, covering issues related to the aesthetics of interaction (e.g. Hassenzahl 2004; Hassenzahl and Monk 2010; Hekkert 2006; Moshagen and Thielsch 2010; Thüring and Mahlke 2007), emotions in technology experiences (e.g. Bødker 2006; Hassenzahl and Tractinsky 2006; Norman 2004) and meaning-making processes in visual technology experiences (e.g. Desmet and Hekkert 2007; Krippendorff 2006; Silvennoinen et al. 2017).

This change in scope has also diversified the variance and complexity of results related to the affective dimensions of HCI. In particular, differences in discussing the underlying dynamics of visual experience in HCI vary greatly. Many hypotheses have been presented to examine the cognitive–affective operations associated with how we make sense of and experience the visual interfaces of technological artefacts. Differing results have been presented, for example, regarding the relationship between aesthetic evaluations and perceived usability. Aesthetically pleasing user interfaces can make us more tolerant of inconsistencies between system properties, affect our attitudes towards technology and positively increase performance (e.g. Norman 2004; Moshagen et al. 2009), but can also have negative effects (e.g. Sonderegger and Sauer 2010). Many of the inconsistencies have emerged due to unsolid theoretical grounds.

Tight industry relationships and a lack of meta-research and replication studies (e.g. Liu et al. 2014) leave aside crucial developments in theory development. This chapter presents an appraisal-theory-based understanding of visual technology experience (Jokinen et al. 2015, 2018; Silvennoinen and Jokinen 2016; Silvennoinen 2017) to clarify theoretical approaches to examining emotional user experience and present methodical possibilities for examining emotions as cognitive processes within HCI.

Affective appraisals are at the core of visual experiences. Technological artefacts are meaningful to the people interacting with them due to the mentally represented qualities people attribute to them (Silvennoinen 2017). The appraisal theory has been used to examine emotional responses in product experiences due to its ability to explain emotion as a process (Demir 2009). Thus, the theory can explain the relationship between visual experience and a design artefact in how a subjective experience emerges from the appraisal process to encounter a design artefact. This further allows the design of such experiments, where the details of the appraisal process can be manipulated to examine the relationship between design artefacts and experiences.

Jokinen et al. (2015) examined the visual experience of shapes using a primed product comparison method. The stimuli for the experiment were pictures of drinking glasses. As the three information sources of the appraisal process have different computational demands (Smith and Kirby 2001), the authors were able to conduct an experiment that examined how these three sources are involved in experiencing product shapes. The primed product comparison method is based on reaction times and preference scores. A participant is given a prime or a cue (i.e. a word) and a stimulus pair (here, two images of drinking glasses). All the different combinations of cues and stimuli are evaluated. The participant is asked to quickly choose between the images based on the word supplied. The speed at which they were required to make the choice enables the detection of different appraisal levels to the time it takes to make a judgment. Experimentally, time is of the essence, as it enables the examination of culturally and linguistically complex elements in conscious experience.

This method can be used to examine the cognitive appraisal process in which subjective experience occurs in considerable detail. Jokinen et al. (2015) identified different levels of the appraisal process in experiencing product shapes, and connected different cues to certain shapes. Some of the appraisal criteria were judged more quickly, indicating that a shorter information processing time is associated with certain appraisal information sources. Stimulus pairs that are dissimilar to each other were also judged faster than pairs with similar shapes. Faster judgments were performed when the appraising cues depicted physical characteristics of the artefacts (e.g. durable or light), which require less associative processing and reasoning than more complex cues (e.g. traditional or timeless).

Silvennoinen and Jokinen (2016) examined appraisals of icons from different design eras. They used the primed product comparison method to examine the process of experiencing icons. Preferences and their processing times were analysed in terms of perceived visual usability and the aesthetic appeal of icons from four different design eras. These two characteristics are underlying dynamics of an overall visual experience (Silvennoinen 2017), which can be further examined in detail with the appraisal theory. Perceived visual usability was operationalised as semantic distance, i.e. the closeness of the icon's pictorial representation to its intended function (e.g. Silvennoinen et al. 2017). The primes or cues were the icons' intended functions (save, print and search). Aesthetic appeal was operationalised using concepts from traditional accounts of aesthetics emphasising positive engagement, intrinsic value and design-era dependency (beautiful, old-fashioned and familiar).

In addition to the differing research results regarding the relationship between aesthetic appeal and perceived usability, ease of interpretation has been reported to enhance aesthetic appeal (Reber et al. 2004; Reppa and McDougall 2015) via cognitive processing fluency; it is reportedly enhanced by stimulus familiarity (Isherwood et al. 2007). Experiencing a visual representation as familiar requires an evaluative appraisal framework. Although an icon would be familiar and easily interpreted, cognitive processing fluency determines whether it is experienced as aesthetically pleasing.

An appraisal-theory-based understanding of visual experience was utilised to explicate the connection between processing fluency, familiarity and aesthetic appeal. Aesthetic appeal and perceived visual usability preferences of the four icons varied, which allowed the icon experience to be examined as a process. Experiencing a stimulus as appealing results from cognitive information processing. Examining the influence of design eras on visual experience, perceived visual usability functions as an underlying factor of aesthetic appeal, and can increase the possibility that an icon will be experienced as pleasing. In addition, judgments regarding visual usability are more unanimous than those related to aesthetic appeal.

Since familiarity increases the fluency of cognitive information processing, familiar stimuli are experienced faster than non-familiar ones. Familiar stimuli have lower activation thresholds for relevant long-term memory nodes, which enables more automated reasoning. However, speed and ease of interpretation do not determine aesthetic appeal. Ease of interpretation is desirable in itself (especially in HCI, where the goals are often efficiency related), but should not be extended to explicate the affective qualities of visual experiences. Visual experience is more complex, as the icon experiment indicates. We are able to interpret familiar stimuli quickly as understandable, but also as unappealing.

Designing for certain targeted experiences is a difficult task. How we experience visual representations varies, and is influenced by numerous factors. There are no predetermined relationships between design elements and how these are experienced. Jokinen et al. (2018) employed appraisal theory and predictive brain theory (Clark 2013) to examine the visual experiences of user interfaces for targeted emotional outcomes. Predictive brain theory integrates recent advancements in cognitive science, indicating that brains constantly match sensory inputs with top-down expectations that support perception and action (Clark 2013).

The primed product comparison method has been combined with eye-tracking measurements to connect certain visual user interface design decisions with targeted experience goals (Jokinen et al. 2018). Participants appraised website designs with affective experiential adjectives, such as modern, civilised and beautiful. This study described the relationship between the three appraisal information sources not as distinct and separate entities, but as interwoven in a complex process. In the appraisal process, bottom-up perceptual stimuli are integrated with top-down associative information and reasoning. Thus, appraisal theory and predictive processing theory can be used to examine the complex relationship between design artefacts and human experience, as they predict how experience occurs in the design appraisal process. The details of the appraisal process can be experimentally manipulated to examine the relationship between design and experience. The methodological approach that combines appraisal theory and predictive brain theory with visual experience can be further utilised to empirically test predictions, for example, concerning cognitive processing fluency and the relationship between visual designs and visual user experience.

3.3.2 Emotional User Experience

In most HCI tasks, the users can be assumed to have goals that they attempt to achieve during the interaction. Given the goal-oriented nature of interactive behaviour, users' emotions during the interaction can be analysed in terms of goal congruency (Jokinen 2015a). Under appraisal theory, the user evaluates the significance of the interaction events to his or her goals. Generally, if the event is appraised as goal congruent, the resulting emotion is positive, whereas goal-incongruent events cause negative emotions. This analysis can be likened to the original intuition of appraisal as an evaluation of an event as either approachable (the event facilitates goal attainment) or aversible (the event obstructs goal attainment).

This level of analysis is still, however, quite trivial, as it does not explain the variety of emotions—positive and negative—that users may experience when interacting with technology. A goal-incongruent event may result in being either frustrated or sad, for instance, and it is the task of the appraisal theory to explain what conditions explain these contingencies: the goal of the appraisal-theory-based analysis of users' emotions is to give a causal account of their emotional responses. For example, Jokinen (2015a) and Saariluoma and Jokinen (2014, 2015) investigated users' subjective emotional responses to task events in computerised tasks. Using a questionnaire with a number of emotion-related items to probe the emotions, and different experiment manipulations, they were able to make causal inferences about the appraisal process. The authors used a competence–frustration model of emotion, where they defined competence as the positive emotion resulting from goal-congruent events such that the user perceives that these events were a result of his or her own actions. Frustration is defined as resulting from goal-incongruent events that obstruct the user from reaching attainable goals.

Saariluoma and Jokinen (2015) tested the goal-oriented nature of competence and frustration. They separated participants into two groups, which either looked at screenshots of online shops or conducted ordinary tasks on them. Then they reported their experiences using an emotion questionnaire. Contrary to the authors' initial expectations, the two groups did not differ in levels of self-reported competence. This means that the participants in the group without a goal-oriented task reported feeling competent. Initially, this seems a strange result due to the definition of competence as the appraisal of task-congruent events that result from applying one's skills. If a user passively watches screenshots without any tasks that manipulate the states of the interactive system, they should not feel competent. However, the authors had also asked the participants about their emotions at the start of the experiment before showing them any stimuli. For the group that did not have interactive tasks, there was a correlation between pre-and post-test questionnaire responses, meaning that their feelings of competence and frustration after watching the screenshots were mainly predetermined, and not affected by the experiment tasks. Conversely, the group that conducted tasks did not have this correlation: their competence and frustration depended on task performance.

Saariluoma and Jokinen (2015) concluded that only events that occur in goal-directed interactions impact emotions, whereas those in non-directed, passive interactions do not. Jokinen (2015a) investigated the connection between task events and emotions in more detail. First, he observed the expected correlation between task performance and emotional responses. He also investigated the impact of pre-task self-reports of emotion, and found that the impact of task performance on emotions depended on prior, pre-task emotions. For instance, Jokinen (2015a) probed the participants' self-confidence at the start of the experiment to test whether self-confidence translates into competence during the experiment. Indeed, participants who reported low self-confidence started the experiment tasks with a below-average sense of competence. However, low self-confidence participants who were successful in the tasks reported higher levels of competence than those who started the experiment with more self-confidence. In other words, self-confidence has a negative moderating effect on the relationship between task performance and self-reported competence. This is in line with the appraisal theory's statement that the same event can result in different emotional responses, depending on the subject.

Jokinen (2015a) also investigated the second stage of the appraisal process—coping—in more depth. He explored two coping strategies, problem-solving oriented and emotional coping. He hypothesised that feelings of competence should result from problem-solving-oriented coping alongside goal-congruent events (that are the result of problem-solving), and that feelings of frustration should be moderated by the emotional coping strategy when there are incongruent events. Jokinen (2015a) treated these coping strategies as individual traits that may vary between users. To measure this trait he used a coping questionnaire, which he adapted to the context of technology use. In the experiment, the participants accomplished tasks in different software environments, such as text editing or image manipulation, and reported their emotions. Jokinen (2015a) observed the expected effect of task performance on competence and frustration. He also confirmed the coping trait hypothesis: self-reported coping traits impacted self-reported emotions, demonstrating that emotion is an individual phenomenon that cannot be understood without referring to the subject's internal processes, such as coping.

3.4   Discussion

The appraisal theory of emotion explicates explains emotion as a process. We here demonstrated the plausibility of applying this theory in an examination of technology experiences and an explication of how experience occurs in HCI. The appraisal as a process operates via cognitively evaluating the subjective significance of an event, and occurs on multiple levels, including physiological level, motivational level and subjective experience. We used multiple example studies, grouped into two main categories, to demonstrate how the appraisal theory can be applied in the study of the human visual experience, and in the study of emotions in technology interaction. The examples showcase the theory's usefulness of the appraisal theory in HCI research, particularly as. As observed, the benefit of appraisal theory in this regard is that it defines emotion in terms of events of the task environment, and in terms of as well as a subject's knowledge and goals of a subject. Thus, the cognitive appraisal process integrates these factors and enables the explication of a variety of subjective emotional experiences in its variety, which are conditioned (and, therefore, explainable) by the circumstances in which the emotion was elicited.

For example, the appraisal theory analyses emotion as a partially cognitive phenomenon, in which information from multiple sources is processed during the appraisal (Smith and Kirby 2001). We exemplified how these information sources of subjective experience are intertwined in top-down and bottom-up processes, leading to testable hypotheses regarding how visual experience occurs and dynamically changes during interactions. The same applies to the appraisal theory-based understandings of visual experience, wherein bottom-up and top-down sources of information dynamically influence the subjective experience.

Different appraisal information sources in experience can proceed from the identification of physical qualities associated with abstract meanings involving higher level cognitive reasoning. For example, appraising a material as warm involves a relatively direct process of temperature recognition and touch perception, but appraising a material as timeless entails association and reasoning. The appraisal theory of emotion is not dependent on sensory modalities. Its logic can explicate technology encounters that are induced by and experienced with different sensory modalities (e.g. Silvennoinen et al. 2015). In addition, the primed product comparison method can be utilised to examine experience as a process pertaining to other senses, such as hearing. The primes can thus be sounds, for example. Using this method, different evaluation times between the stimuli and the primes indicated differing mental processes in visual experience, as predicted by appraisal theory.

The methodology presented in this chapter provides grounds for HCI researchers to examine subjective experiences in HCI. Designers can also use the primed product comparison method to analyse how well their visual designs (and the intended experience goals) correspond to users' experiences.

References

Arnold M (1960) Emotion and personality. Columbia University Press, New York

Baddeley A (2007) Working memory, thought, and action. Oxford University Press, Oxford

Barrett LF (2006) Are emotions natural kinds? Perspect Psychol Sci 1(1):28–58

Bradley MM, Lang PJ (1994) Measuring emotion: the self-assessment manikin and the semantic differential. J Behav Ther Exp Psychiatry 25(1):49–59

Bødker S (2006) When second wave HCI meets third wave challenges. In: Proceedings of the 4th nordic conference on human-computer interaction: changing roles. ACM, pp 1–8

Cacioppo JT, Berntson GG, Klein DJ, Poehlmann KM (1997) Psychophysiology of emotion across the life span. Ann Rev Gerontol Geriatr 17:27–74

Cacioppo JT, Berntson GG, Larsen JT, Poehlmann KM, Ito TA (2000) The psychophysiology of emotion. In: Lewis R, Haviland-Jones JM (eds) Handbook of emotions, 2nd edn. Guildford, New York, pp 173–191

Clark A (2013) Whatever next? predictive brains, situated agents, and the future of cognitive science. Behav Brain Sci 36(3):181–204

Demir E, Desmet PM, Hekkert P (2009) Appraisal patterns of emotions in human-product interaction. Int J Des 3(2):41–51

Desmet P, Hekkert P (2007) Framework of product experience. Int J Des 1(1):57–66

Drachen A, Nacke LE, Yannakakis G, Pedersen AL (2010) Correlation between heart rate, electro-dermal activity and player experience in first-person shooter games. In: Proceedings of the 5th ACM SIGGRAPH symposium on video games. ACM, pp 49–54

Ekman P (1992) Are there basic emotions? Psychol Rev 99(3):550–553

Ekman P (1999) Basic emotions. In: Dalgleish T, Power M (eds) Handbook of cognition and emotion. John Wiley & Sons, New York, pp 45–60

Ekman P, Friesen WV (1971) Constants across cultures in the face and emotion. J Pers Soc Psychol 17(2):124–129

Frijda N (1988) The laws of emotion. Am Psychol 43(5):349–358

Hassenzahl M (2004) The interplay of beauty, goodness, and usability in interactive products. Hum Comput Interaction 19(4):319–349

Hassenzahl M, Monk A (2010) The inference of perceived usability from beauty. Hum Comput Interaction 25(3):235–206

Hassenzahl M, Tractinsky N (2006) User experience – a research agenda. Behav Inform Technol 25(2):91–97

Hekkert P (2006) Design aesthetics: principles of pleasure in design. Psychol Sci 48(2):157–172

Isherwood SJ, McDougall SJP, Curry MB (2007) Icon identification in context: the changing role of icon characteristics with user experience. Hum Factors 49:465–476

Izard CE (2007) Basic emotions, natural kinds, emotion schemas, and a new paradigm. Perspect Psychol Sci 2(3):260–280

Jokinen JPP (2015a) Emotional user experience: traits, events, and states. Int J Hum Comput Stud 76:67–77

Jokinen JPP (2015b) User psychology of emotional user experience. Jyväskylä Stud Comput 213

Jokinen JP, Silvennoinen J, Kujala T (2018) Relating experience goals with visual user interface design. Interact Comput 30(5):378–395

Jokinen JP, Silvennoinen JM, Perälä PM, Saariluoma P (2015) Quick affective judgments: validation of a method for primed product comparisons. In: Proceedings of the 33rd annual ACM conference on human factors in computing systems. ACM, pp 2221–2230

Krippendorff K (2006) The semantic turn: a new foundation for design. Taylor & Francis, CRC Press, London, New York, Boca Raton

Lazarus RS (1966) Psychological stress and the coping process. McGraw-Hill, New York

Lichtenstein A, Oehme A, Kupschick S, Jürgensohn T (2008) Comparing two emotion models for deriving affective states from physiological data. In: Peter C, Beale R (eds) Affect and emotion in human-computer interaction. Springer, Berlin, Heidelberg, pp 35–50

Liu Y, Goncalves J, Ferreira D, Xiao B, Hosio S, Kostakos V (2014) CHI 1994–2013: mapping two decades of intellectual progress through co-word analysis. In: Proceedings of the sigchi conference on human factors in computing systems. ACM, pp 3553–3562

Mandryk RL, Inkpen KM, Calvert TW (2006) Using psychophysiological techniques to measure user experience with entertainment technologies. Behav Inform Technol 25(2):141–158

Moshagen M, Musch J, Göritz AS (2009) A blessing, not a curse: experimental evidence for beneficial effects on visual aesthetics on performance. Ergonomics 52(10):1311–1320

Moshagen M, Thielch MT (2010) Facets of visual aesthetics. Int J Hum Comput Stud 68(100):689–709

Norman D (2004) Emotional design. Basic Books, New York

Partala T, Surakka V (2003) Pupil size variation as an indication of affective processing. Int J Hum Comput Stud 59(1–2):185–198

Picard RW (1997) Affective computing. MIT Press, Cambridge, MA

Reber R, Schwarz N, Winkielman P (2004) Processing fluency and aesthetic pleasure: is beauty in the perceiver's processing experience? Pers Soc Psychol Rev 8(4):364–382

Reppa I, McDougall S (2015) When the going gets tough the beautiful get going: aesthetic appeal facilitates task performance. Psychon Bull Rev 22(5):1243–1254

Russell JA (1980) A circumplex model of affect. J Pers Soc Psychol 39(6):1161–1178

Russell JA (2009) Emotion, core affect, and psychological construction. Cogn Emot 23(7):1259–1283

Russell JA, Barrett LF (1999) Core affect, prototypical emotional episodes, and other things called emotion: dissecting the elephant. J Pers Soc Psychol 76(5):805–819

Saariluoma P, Jokinen JP (2014) Emotional dimensions of user experience: a user psychological analysis. Int J of Hum Comput Interaction 30(4):303–320

Saariluoma P, Jokinen JP (2015) Appraisal and mental contents in human-technology interaction. Int J Technol Human Interact 11(2):1–32

Scherer KR (2009) The dynamic architecture of emotion: Evidence for the component process model. Cogn Emot 23(7):1307–1351

Shiota MN, Campos B, Oveis C, Hertenstein MJ, Simon-Thomas E, Keltner D (2017) Beyond happiness: building a science of discrete positive emotions. Am Psychol 72(7):617

Silvennoinen J (2017) Apperceiving visual elements in human-technology interaction design (Doctoral dissertation). University of Jyväskylä Press, Jyväskylä

Silvennoinen JM, Jokinen JPP (2016) Aesthetic appeal and visual usability in four icon design eras. In: Proceedings of the 2016 chi conference on human factors in computing systems. ACM, pp 4390–4400

Silvennoinen JM, Kujala T, Jokinen JPP (2017a) Semantic distance as a critical factor in icon design for in-car infotainment systems. Appl Ergon 65:369–381

Silvennoinen JM, Rousi R, Jokinen JPP, Perälä PM (2015) Apperception as a multisensory process in material experience. In proceedings of the 19th international academic mindtrek conference. ACM, pp 144–151

Silvennoinen J, Rousi R, Mononen L (2017b) Creative interpretation in web design experience. Des J 20(sup1):S134–S145

Smith CA, Kirby LD (2001) Toward delivering on the promise of appraisal theory. In: Scherer KR, Schorr A, Johnstone T (eds) Appraisal process in emotion. Oxford University Press, New York, pp 121–138

Sonderegger A, Sauer J (2010) The influence of design aesthetics in usability testing: effects on user performance and perceived usability. Applied Ergonomics 41(3):403–410

Thüring M, Mahlke S (2007) Usability, aesthetics and emotions in human–technology interaction. Int J Psychol 42(4):253–264

Zimmermann P, Gomez P, Danuser B, Schär S (2006) Extending usability: putting affect into the user-experience. In: Proceedings of nordichi'06, 27–32

Part II
Design

Design is a cognitive action, both mental and embodied that forms the core of human thought in innovation and technological creation. Emotional design problems deserve specific analysis that focuses on these cognitive actions. They may be explicitly observed in scientific research probing such as thinking aloud, or even through embodiment processes such as sketching and implementation of design tools (customer journeys, business model canvases etc.). Emotions are always attached to these cognitive actions. They guide the direction of the design, aid in setting priorities and hierarchies, and contribute to the understanding of how to create what is learnable and memorable within the designs. In other words, human actions are always decided upon the grounds of emotional states. This determines how design professionals, customers, users and onlookers rank the importance of an event, technical artefact, service or even person (e.g., the person or people attached to the design—its creator, its endorser and more importantly its consumer). Emotions may appear irrelevant in the midst of technological design problem-solving, yet if ignored, almost always contribute to greater problems in the human-technology interactions to follow.

It is therefore, crucial to connect the elements, actions and processes of design and development to emotions on the individual, group, community and societal levels. Designers not only contribute to the augmentation, advancement and manipulation of technology, they subsequently play a role in shaping our world and how we emotionally experience it based on relationships—organization and hierarchies, manipulation of physical space and action, mediation of human-to-human relationships and so forth. Design, and indeed its consumption, can be related to John F. Kennedy's notorious statement, "Ask not what your country can do for you—ask what you can do for your country". Thus, in the emotional experience of design multi-layered, circular and reciprocal relationships exist between the designers, producers, consumers and onlookers that see all as active participants within this dynamic network of interactions. This current part is devoted to chapters concentrating on design, design tools and methodologies and designing innovation for specific emotional states.

Chapter 4
Research for Designing for Emotions

Gilbert Cockton

Abstract Designing for emotion is very well established in older design disciplines such as advertising, fashion, graphic, interior and product design, and has made good progress in more recent disciplines such as service design. Design work integrates a range of creative practices with technical and business considerations. There is a continuum of practices from highly demanding technology design to highly creative craft work. Even at the most technical extreme, there are vital creative aspects, especially when faced with challenges that are novel for an area of design, such as designing for emotional aspects of technology usage. The outcomes of novel creative practices cannot be known in advance, making it impossible to fully plan creative work. Researching *for* design—that is, in support of truly novel creative design work—must carefully consider the work practices of the designers who are expected to benefit from new theories, knowledge, information, processes or procedures. Any research on emotions that assumes good science can be automatically applied to any design context will severely limit its potential impact and reach. It is unlikely to be attractive research that can be realistically applied. This chapter begins with the consideration of how emotions are considered in design, summarises the history of an affective focus in human–computer interaction and reviews the author's practical support for an affective focus within interaction design. It then relates these three reviews to some key perspectives from creative design research to support an initial agenda for effective impact for research for affective design.

G. Cockton (✉)
School of Computer Science, University of Sunderland, Sunderland, UK
e-mail: gilbert.cockton@sunderland.ac.uk

© Springer Nature Switzerland AG 2020
R. Rousi et al. (eds.), *Emotions in Technology Design: From Experience
to Ethics*, Human–Computer Interaction Series,
https://doi.org/10.1007/978-3-030-53483-7_4

4.1 Do Designers Need to Learn About Emotions, or Do Emotion Researchers Need to Learn About Designers and Design Research?

Design is an extremely broad field, ranging from designer-makers' 'truth to materials' (Moore 1934) to highly analytical science-based work in engineering. At one extreme, designer-makers 'feel' what is in the materials and the aesthetic and semiotic opportunities that this offers. At the other extreme, systematic engineers focus on meeting requirements specifications by demonstrably solving a clearly stated problem. The apt question here is thus: which designers need to learn about emotion?

Most design teams have no burning need for resources or new approaches to designing for emotion, which has been routine for millennia. Spectacles such as processions, animal parades, gladiator battles and chariot races were commonplace in the Roman Empire and for centuries before that, and not only in the most advanced civilisations. These events were out of the ordinary, wastefully extravagant, entertaining and meaningful. They all sought to evoke emotional responses from the audience. Several design disciplines were involved in the creation of textiles, costumes, vehicles, interiors and exteriors, and the smaller artefacts that were vital to a spectacle.

Fashion designers, games designers, transportation designers, interior designers, film makers, animators and architects perform similar work today. Furniture designers, product designers and graphic designers would also recognise many of the creative aesthetic craft practices from past millennia. The ability to spark emotions is central to the craft of more recent disciplines that combine design and media, such as advertising. More recently, service design also aims to create appropriate customer experiences across a range of physical and digital touch points, potentially combining interior architecture, workwear design, branding, contact centre scripting and digital design.

Interaction design struggles with emotions due to the spread of disciplines involved, with science and engineering more at home with instrumental rather than experiential values. Given the major shortage of studio-educated interaction designers, long-established software and hardware suppliers have had engineers design user interfaces for decades, perhaps with input from ergonomists. Unlike design disciplines that draw exclusively from studio-educated graduates, interaction designers have diverse backgrounds that make it difficult to predict their competence and ability to design for emotion. In contrast, a graphic designer or copywriter could not work on branding, nor an interior designer on restaurant design, if they did not understand how to shape a range of meanings and experiences, focusing on emotions as and when necessary. Their expertise in this context would be craft based, with little or no explicit science. They would typically be aware of concepts such as semiotics, but the author of a fragrance strapline 'weapon of mass seduction' may not consciously derive it as a Saussurean paradigmatic substitution. Such wordplay is native to a copywriter's craft.

It should be clear to anyone who visits a decorative arts or transport museum that designers have evoked emotions in consumers and voyeurs for millennia. There is no pervasive problem with emotions across design. Where emotions are poorly considered, this is typically due to instrumentally focused technologists who want to meet specified requirements to solve a clearly specified problem. However, any poor consideration of emotion here, such as an 'Abort now?' message to users who could include pregnant women, is part of a much broader problem: the lack of appropriate human foci in mainstream engineering design. Better consideration of emotion in engineering design cannot be achieved in isolation. A more comprehensive range of human factors needs to be considered and smoothly integrated into design work. Impoverished engineering design processes must be replaced with more balanced, integrated and generous practices (Cockton 2020a).

A belief that psychological knowledge of emotions can improve creative design practice is consistent with engineering design thinking that is rational, idealised and linear ('RILED', Cockton 2020a). Knowledge is seen as a solution to a problem, which must be identified first. Once the problem is well specified, a linear process is followed: the problem is solved by a validated solution. This process is rational since rationality demands that a problem be clearly stated before attempting to solve it. This valuing of upfront plans is millennia old, starting with Ancient Greek philosophy and perhaps earlier, and becoming dominant with the rationalists of the Scientific Revolution (Gedenryd 1998). Like the final version of a mathematical proof, reason must proceed in a linear manner, demonstrating a systematic progression from premises to conclusions. Unfortunately, such a belief in upfront re-usable processes has no basis in fact in any empirical study of creative innovation work (Cockton 2020a). Fixed upfront design processes only work for 'tame' problems and not the 'wicked' ones that dominate creative work (Rittel and Webber 1973).

It is thus the case that *some* designers could better consider emotions in their work, but equally researchers could learn from many established designers how emotional meanings are created through form and content. If scientific knowledge of emotions is to improve design work, then a broad understanding of creative design is required for a balanced range of options to be considered. One set of practices that may be problematic for affective design is the instrumental problem-solving of idealised engineering design. However, this is only one approach to design work, which only works for (inappropriately) tame(d) problems. One set of practices that may be problematic for simple minded research is the continuous problem setting with its co-evolution of problems and solutions (Schön 1983), which is marginalised in idealised engineering design. Rather than work on problems then solutions in a strict linear sequence, both are worked on concurrently, co-evolving through backtalk from 'the materials of a design situation' (Schön 1992). This is more common than the linear ideals of many engineering design text books and standards. A process remains linear even with iteration if there is no concurrent progression of different aspects of design work.

Different design progressions have different implications for research for design. In a linear design process, scientific knowledge of emotions could be provided in declarative and procedural forms prior to projects, and then systematically applied at the right points in a fixed process. Such an approach can use a single mode of design research (research *for* design) to develop novel approaches and resources. However, this is not possible for concurrent design work where affective design resources could be repeatedly reconsidered throughout a design project. To provide effective support for such concurrent design work, researchers need additional competence in, or at least knowledge of, three modes of design research, as outlined in (Frayling 1993): research *into* design, research *through* design and research *for* design.

Researchers who develop novel resources and approaches for affective design thus need to understand a wide range of approaches to both design work *and* design research. For interaction and software design, researchers may incorrectly assume that designers follow textbook or standard methods such as ISO-9241-210 (ISO 2019) or similar. When combined with a belief that all designers need help to design for emotions, researchers will severely limit their potential impact and effectiveness. If the full range of design research approaches are ignored, then the chances of failure significantly increase.

The first of Frayling's (1993) modes of design research is research *into* design, which studies creative work and its outputs in the past and present. This includes evaluating innovative design practices. The second, research *through* design, uses creative design practices as the backbone of research methodologies that create new knowledge by developing new artefacts, approaches, resources, and work progressions. The third, research *for* design, aims to improve creative design work by making it better informed, both declaratively and procedurally, that is, with knowing *that* and knowing *how*. Frayling (1993) also distinguished between what he referred to as 'Big R' research (academic, original) and 'little r' research (pragmatic, routine). Most design work involves the latter: it informs a project but is not expected to be an original contribution to a broader body of knowledge.

All three modes of design research are needed to develop new approaches and resources for design work. Research *into* design needs to draw on both secondary and primary sources to avoid unpleasant surprises. Researchers who aim to support creative work need to be knowledgeable about how designers work, from the extensive literature on creative practices. Any proposed 'improvement' to creative design work must have a clear value proposition that will be plausible to open-minded reflective designers.

Providing support for a combination of concurrent dynamic problem-setting and progressive development of solutions requires all three modes of design research. To better support consideration of emotions in design, research *through* design is required to try out novel approaches and resources developed by research for design in realistic contexts. Research *into* design (here into the progression of research through design) is needed to evaluate the impact of novel approaches and resources.

Researchers who proceed immediately and exclusively to research *for* design reduce their chance of being impactful. They need instead to add research *into* and *through* design, testing out new approaches and resources on creative projects before disseminating. It is important to develop new approaches and resources in realistic contexts, and to evaluate their effectiveness through independent research *into* design. The next section gives examples of how I took this approach with worth mapping resources at Microsoft Research Centre Cambridge (Cockton et al. 2009a), with their effectiveness evaluated by independent research *into* design in the VALU project (Cockton et al. 2009b).

Such multiple research modes provide the required breadth for careful studies of design teams' positive and negative experiences with novel research contributions. There is no attempt to check whether novel approaches and resources are used exactly as expected. Replication and creative work do not mix well. This requires us to consider a third form of design practice, in addition to problem-solving and problem-setting: design work can be *exploratory*. Research for design here is experimental, investigating opportunities in an open manner. Much professional practice takes this approach, trying things out and reflecting on their worth. This is how emotions were first considered in human–computer interaction (HCI). A chicane followed that squeezed research into a single focus. The scope of work on emotions in HCI then expanded. The next section considers this chicane and the entry to and exit from it.

4.2 The Emotion Chicane in Human–Computer Interaction Research

Research on the emotional aspect of HCI started broad, narrowed down to automatic recognition of emotions, and then broadened out to a rich consideration of the many ways in which emotions manifest when interacting with computers.

User experience (UX) was initially a practical focus that was understood as an extension of usability. It was originally rooted in industrial research and development (R&D). Usage of the UX term began several years after usability became a founding concern of HCI research: a 'focus on user experience … expanded and changed our view of usability engineering. … Qualities of usability such as productivity, enthusiasm, and control emerge from the user's interpretation of, and response to, a computer system while engaged in everyday work' (Whiteside et al. 1988, p. 813). The focus gave access to 'the richest possible data on user experience … collected within their natural work context. … time, motivational, and social contexts will differ, and a variety of products will be used in parallel' (Whiteside et al. 1988, p. 806). Cognition and affect are not separated in this account. Whatever is witnessed in a user's experience is potentially relevant. Whiteside and colleagues' account of the evolution of usability engineering at IBM and DEC was an early example of what has been called 'second-wave' HCI and its 'turn to the social' (Rogers et al. 1994).

A few years after this account was published, a different research group at IBM worked on a system for EXPO'92 in Seville. A move into non-work settings brought them into contact with laughter, quality of life, empowerment and lingering to learn (Gould et al. 1997). Again, there is no separation of cognition and affect here. UX was whatever a user experienced; it did not require any theoretical foci to dissect it into any academic disciplinary categories. UX and associated feelings were seen as providing a better basis for understanding usability. Cognitive, affective and other factors intertwine in initial HCI accounts and understandings of UX.

A decade later, this intertwining gave way to a predominantly affective focus, for example, the Microsoft Desirability Toolkit (originally *Product Reaction Cards*, Benedek and Milner 2002), an R&D development of 118 cards with single words or phrases that users could select during an evaluation to indicate how they felt when using a product. Example cards included stressful, gets in the way, patronizing, time-consuming, intimidating, inviting, confusing, sophisticated, frustrating, fun, and exciting. This focus contrasts strongly with the initial atheoretical focus on UX and everything that came with it, which made a balanced approach possible.

The editorial of a journal special issue on emotion and HCI observed, 'people have always been touched by some aspects of computers—from teletype print outs of Snoopy in the 1970s to the empathic communities of the 1990s internet, and from the first interactive computer games to the brand-intensive websites of the e-business revolution' (Cockton 2002). This special issue had a narrow focus on systems that adapted to users' emotional states, an extreme of specialisation for affective HCI. By seeking to 'recognise' a user's emotions, researchers committed themselves to a theory of discrete named emotions, rather than a less fraught dimensional theory with a continuous space with dimensions for valence (positive to negative) and arousal (strength of feeling). HCI research on emotions had entered a chicane after initially following a broad road. Affect had been separated out as a specialist area, and was often reduced to something to be detected and ameliorated when necessary (Boehner et al. 2007).

A follow-up journal issue (Cockton 2004) published a range of commentaries on the tightly focused papers in the special issue. Four commentaries were broadly sympathetic to the special issue's research on adapting interaction to detected emotions, but recommended refining it to deal with a range of issues related to psychological theory, feasibility, physiological measures and instruments, and ethics. Other commentaries were less sympathetic. One questioned the assumption that reducing negative emotions is a universal design goal, arguing that intelligent adaptation can de-skill and reduce trust. Another commentary focused on the relationship between language and emotional response; it introduced media authoring as an alternative to 'intelligent' algorithms, which cause unintended emotional responses when they are clumsy. In contrast, users can reinterpret clumsy authoring in media spaces. A classic human factors design issue is in scope here: do we leave the management of emotion to the computer or do we share it with users? As Boehner et al. (2007) had noted, the pendulum may have swung too far towards automation, away from a more democratic and respectful partnership between human and machine.

Two other commentaries focused on how media authoring (including interface text and graphics) offers more proven opportunities to co-influence users' emotions than those provided by attempted one-sided control through speculative adaptive algorithms. There was an unusual example of a medium in the closing commentary, *mildew*. We would expect this to provoke negative reactions: interactive resolution of adverse emotions would not be appropriate here. This suggests that a comprehensive understanding of the role of emotions in interaction is needed before we can make well-grounded design interventions, algorithmic, authored or otherwise. Meaning has an important role here, as it bridges the gap between simple and complex emotions to connect interaction design with longer established design practices in fashion, transport, graphics and interiors, where desired emotional responses frame the direction for much design intent.

With this return to a broader range of emotions, HCI exited a chicane by establishing the need for a stronger and broader theoretical framework. Several frameworks were now under consideration, ranging from the algorithmically convenient (e.g., discrete emotion theory) to critical perspectives on emotion, media and meaning. As the chicane widened, Boehner et al. (2007) critically surveyed the range of constructs, theory and assumptions in affective HCI research, arguing for 'emotion as interaction: dynamic, culturally mediated, and socially constructed and experienced'. This critique, part of the rich breadth of third wave HCI, confronted simple recognition of discrete emotions with perspectives on embodiment (physiology), culture (language and media), meaning (self-reporting, complex emotions) and axiology (ethics). The possible understandings of what emotions could be, and how they manifest themselves, were much broader than in initial industrial R&D conceptualisations of UX. I made some use of this broadening in my own research, as described in the next section.

4.3 Practical Support for Considering Emotion in Interaction Design

In this new broader critical context, I found myself with a very practical need. In 2007, I was seconded to Microsoft Research Cambridge (MSRC) to develop approaches to explicitly consider *worth* in IxD. Worth is a broad concept that subsumes value, values and experience by considering the balance of positives and negatives across UX and usage outcomes (Cockton 2006). As with all valenced axiological considerations, emotions are involved.

From a design perspective, it is helpful to be explicit about the relationship between a digital artefact and the experiences and outcomes that result from its usage. Some marketing and advertising approaches have connected products to values. I adapted the main resource for one of these approaches, the hierarchical value model (HVM), for use in IxD (Cockton et al. 2009a). An HVM links product attributes to usage consequences and outcomes through simple unidirectional linear means-end chains (MECs). The underlying theory here is that product advertising and market positioning that explicitly connect with consumers' values will improve their willingness to consider a product. In an HVM, emotions are considered as psychosocial *consequences* of ownership and usage that are means to the higher ends of consumers' values. Emotions in this context are not ends in themselves but means to ends. In such a framing, designing for emotions seeks to foster experiences that can deliver on consumers' values.

There were known issues with HVMs in marketing research: MECs could become complicated, bidirectional and nonlinear. Nevertheless, I adapted HVMs, working first with the TEKES VALU project (Cockton et al. 2009b) and then with colleagues at MSRC, when the adapted HVMs were called *worth maps* (Cockton et al. 2009a). MEC structures were adapted as both projects progressed. At MSRC, I modularised the complexity arising from usage consequences with a separate tabular representation of a UX. This made worth maps more manageable. There were three drivers here. The first was prior knowledge that HVMs *could* become too complicated. The second driver was that worth maps in practice *had* become too complicated when modelling MECs for a digital family archive that was being developed as a research prototype. The third 'driver' was adventitious. As part of a research through design methodology, a living document was maintained that tracked the evolution of different aspects of the design space. The section on usage consequences had a table format that was very similar to that for use case tables from software engineering, so I extended this existing software engineering representation of usage to simplify worth maps. This was an example of a 'reflective conversation with the materials of a design situation' (Schön 1992). Documentation of co-evolving problem and solution spaces initiated backtalk that suggested that table formats could be used to move complexity out of the middle of MECs in worth maps. Tables for the consequences in a worth map would allow considerable simplification by moving the detail of an interaction to a related design resource.

This new table resource was originally called a UX Frame (Cockton 2009), but is now called a UX Case (UXC), realigning it with the use case structure on which it is based (Cockton 2020b). This is an example of an exploratory approach (Sect. 4.1). A problem was being addressed (HVM complexity), but a possible solution emerged fortuitously that turned out to 'solve' a wider set of problems. The details of worth maps do not matter for this chapter. They can be found elsewhere (Cockton 2009, 2020b; Cockton et al. 2009a, b). Only UXCs are relevant as, values apart, they removed all consideration of emotion to a separate resource that can be used with or without worth maps.

The core of a UXC is a two-column use case from software engineering (Table 4.1, Cockburn 2000). Use cases focus on functionality, with occasional elements of cognition. The left-hand column in Table 4.1 thus mostly shows user input actions with some cognitive operations (e.g., Cell 27: review, determine that like, choose to add). The right column shows system actions.

I extended this format with additional columns to cover a range of UX considerations. Up to four additional columns can be added to this use case format to accommodate the broad concerns of third wave HCI:

- Feelings
- Beliefs
- Social interactions
- Physical actions.

With these additional columns, we can envisage how feelings and impressions foster worthwhile experiences and avert adverse ones. UXCs can also be used to anticipate usage difficulties by exposing interactions that may be unpleasant, inefficient or even come to premature unsatisfactory ends. UXCs let designers expose

Table 4.1 E-commerce transaction use case excerpt (Cockburn 2002, p. 95)

User actions	System responses
20. Shopper will select a product model	21. System will determine standard product model options, and then present the first question about determining major product options
22. While questions exist to determine product option recommendations	24. System will prompt with questions that vary based on previous answers to determine the shopper's needs and interests related to major product options, along with pertinent information such as production information, features and benefits, comparison information, and pricing
23. Shopper will answer questions	
25. Shopper answers last question	26. At the last question about major product option desires, the system will present the selected model and selected options for shopper validation
27. Shopper reviews their product selection, determines they like it, and chooses to add the product selection to their shopping cart	28. System will add product selection and storyboard information (navigation and answers) to the shopping cart
	29. The system presents a view of the shopping cart and all of the product selections within it

and assess the likely path of a UX by indicating how feelings, beliefs, and social and physical actions are expected to arise and steer subsequent interactions. UXCs articulate how a UX falls into place as an interaction progresses to satisfy instrumental goals and achieve some terminal purpose. This supports a detailed approach to storymapping (Patton with Economy 2014).

Table 4.2 shows a UXC for booking a van via an imaginary lovelyvan.com website. A dotted arrow traces the van booking experience from top to bottom. There are five rather than six possible columns, as its rightmost 'Actions in the World' column combines social and physical actions. There is a full textual expansion of this UXC, with extensive details of the interaction, as a worth delivery scenario in Cockton (2020b).

The interaction begins at the top with Sally and Harry deciding to try a van hire website recommended by friends. They work collaboratively and plan as they go, establishing that a hire depot for LovelyVan is near enough to their home for it to be worth hiring a van from them. After selecting the nearby depot, there is an impressive animated transition ('that's cool') with sufficient detail provided to let them make a confident choice and book a van after checking that all their details are correct. They are impressed by the booking email and print out its attachment and pin it up.

In Table 4.2, the achieved purpose is instrumental, since booking a van is only a means to an end. All that is achieved in this UXC is confidence in a good plan that is partially implemented by the booking. The actual purpose, that is, the terminal value, of hiring the van is moving objects between two or more places. When ending with such instrumental rather than terminal benefits, the final feelings should be positive, as in Table 4.2.

Table 4.2 Good plan for van hire UXC (Cockton 2009)

Feelings	Beliefs	User action	System response	Actions in the world
Worth a try		Open lovelyvan.com	Display homepage	
Not a good place to start	Can find prices and availability			
				Sally persuades Harry
		Enter post code		
			Show depots map	
	Nearest depot is on ring road	Sally sees nearest depot		
		Select depot on ring road	Display depot and van info	
That's cool				
	Can find right van	Select appropriate van	Display book this van page	
				Sally checks details
		Book and pay for van		
			Display and email confirmation	
Feels great, all well planned		Save and print confirmation page		
	Booked right van for right time period			
		Read email, follow link to info pdf	Display pdf	
That looks very smart				
		Print info and instructions pdf		Staple and pin up info and instructions
Looking forward to getting van	Have all necessary details			

As well as solving a problem related to worth maps for the family archive prototype at MSRC, this form of representation inspired an interaction designer to develop a much more extensive usage journey illustration several months later with a colleague, further demonstrating multiple unanticipated benefits from exploratory approaches to design. The above table format was also used successfully in a Master's project to develop a new worth-focused information system for the Finnish Golf Union (Vu 2013).

There was limited use of existing academic research in the development of worth maps and UXCs. The latter were based on software design practice, combined with an understanding of the expanding scope of HCI research across its three waves. Worth maps were based on a marketing resource, the HVM, which in turn was based on simple motivational theory and values research from social psychology. I did not make full use of every theory on emotions that I had become aware of through editing two special issues (Cockton 2002; 2004), for which I had read some of the key background literature. I also supervised two Ph.D.s on brain–computer interfaces (Doherty et al. 2002, Gnanayutham and Cockton 2008) and was familiar with physiological measures. However, little of this was relevant to designers reflecting on the role of feelings during interaction (Cockton 2009). What was more useful was the extensive literature on values from social psychology (Cockton 2020b).

The main point here is that I had no prior commitment to applying any specific psychological research on, or theories of, emotion. Instead, I applied a broader pragmatic axiological perspective that considered all valenced human phenomena, for example, emotions, appraisal, utility, motivation, or values. To have decided in advance that some psychological research or theory was going to be relevant and applicable would not have been *research for design*. Instead, it would have been applied psychology that focused on a small set of factors. However, design must be broad and holistic in its consideration.

4.4 Informed Resourcing of Design Work

The previous sections span over four decades of HCI practice in industry and research, moving from broad reviews to a specific focus on my own research on when and why feelings matter in research for affective interaction design. The development of novel practical approaches and resources for design work needs to carefully consider the wide range of issues that were identified in the first three sections. These issues can be grouped as follows:

- The knowledge, experience and expertise required from a design team
- The expected working practices of design teams
- The effort required to learn and initially use new approaches and resources.

For the first group of issues, there is a risk in thinking that a design team will have the same disciplinary range as a research team. Research teams often have project-specific multidisciplinary expertise that can identify theories to apply, and when and how to apply them. For example, the research team for *Communicating the Experience of Chronic Pain and Illness Through Blogging* (Ressler et al. 2012) comprised a specialist nurse, an emergency medicine doctor, an HCI specialist and an epidemiologist. Similarly, the research team for *Design to Promote Mindfulness Practice and Sense of Self for Vulnerable women in Secure Hospital Services* (Thieme et al. 2016) included mental healthcare professionals. It is unwise to assume that if a research team can do something then a typical design team can. Even if they can, for example, through specialist expertise from their client, they may not follow 'the same' approach as the research. This leads into the second group of issues related to expected working practices.

It is important to understand what design teams already do well, as well as what can and cannot transfer from research settings. Design teams differ, so it is important to be clear on assumptions about what designers know and can do. While most studio-educated creative designers will have extensive craft skills for designing for emotion, this is less likely among interaction designers with STEM (science, technology, engineering and mathematics) backgrounds. Even if scientific breakthroughs offer radically new effective approaches for experienced creative designers, this new knowledge must be framed, presented and archived in ways that fit existing creative work patterns, and be compatible with how creative designers learn and apply new knowledge.

No wholesale changes to the ways in which design teams work should be required to take advantage of new resources from design research. The 1960s design methods movement made this mistake and quickly lost momentum, in part because one of its founders, John Chris Jones, later acknowledged that:

> I didn't want to get involved with design theory or methods ... I did this ergonomic study of how the designing was done purely with the view of getting the ergonomic information, which was obviously sound and well tested, into the engineering design process at the point where it wouldn't be rejected, ... in doing that I hit on what's now called design methods. (Mitchell 1992)

Human factors ('ergonomic information') constitute only one set of factors in design. There are many others. Rebuilding a design process around a single 'centre' is a poor formula for success, which Jones (1988) later acknowledged. Fifty years after the failure of the design methods movement, we need to stop all attempts to completely program creative design work. Any attempt to impose method or process will fail. This will be not only due to resistance from design teams, but also it is almost always impossible because few approaches and resources are complete enough to deliver repeatable methods. Too much relevant knowledge is project-specific to allow off-the-shelf shrink wrapped methods (Woolrych et al. 2011). Design teams have to work to get methods and processes to work. Researchers cannot control them from a distance.

Academic research varies in its sensitivity to actual creative design practices. At its most insensitive, there is a naïve belief in the power of scientific knowledge—that designers will welcome enlightenment with open arms and completely change work routines to accommodate it. We need realistic approaches to bringing the benefits of research to design. Such realism needs to consider the third group of issues above.

The effort required to learn and initially use new approaches and resources needs to be carefully addressed. A move from processes and methods to approaches and resources should focus researchers on how design teams can learn new practices. Resource-based approaches such as Business Model Canvas (Osterwalder and Pigneur 2010) and Value Proposition Canvas (Osterwalder et al. 2004) have fared well in comparison to the results of much HCI research, as have Personas (Pruitt and Adlin 2005). What these resources and supporting approaches have in common is a flexibility that lets design teams adapt them for specific projects. All have involved community crowdsourcing of examples and have not given priority to applying specific scientific knowledge or theories, although Value Proposition Canvas does build on *Jobs To Be Done* (Kalbach 2019) from management research.

Resource-based approaches do not attempt to fully direct design teams. Instead, they provide public resources such as templates, canvases, information sources, diagram formats, walkthrough practices and guidance that improve design teams' competences without ineptly constraining their work.

As with any human-focused activity, research for design needs to be based on appropriate understandings of its beneficiaries and respect for their knowledge, values, and ways of working. Design practice innovations become widely adopted and effective when they deliver value in work settings—not because the underlying science adheres to the academic disciplinary values of funders, reviewers and editors. An agenda for impactful research for designing for emotions needs to start with this recognition and build on it to improve the chances of effective resourcing for creative design practice.

4.5   An Agenda for Impactful Research for Designing for Emotions

To be relevant to design practice, research on the role of emotions in design needs to be compatible with design work, as set out in the previous section.

Research needs to be adapted to design practice (the reverse won't happen). This chapter aims to provide a broad perspective on creative design practice. It does not consider tame problems (Rittel and Webber 1973) since these are understood well enough to not require design practice innovation. In fact, they can be so routine that they are simply work tasks requiring little creative practice. An initial agenda based on this broad perspective is as follows. A more detailed analysis can be found in (Cockton 2020a).

The agenda outlines an implicit methodology for research for design projects that blends: (1) understanding creative design practice by developing novel approaches and resources on the basis of realistic expectations and genuine commitments; and (2) collecting and sharing examples of use and guidance. The agenda gives rise to three overlapping sets of activities for research for design.

4.5.1   Understanding Creative Design Practice

There have been over 50 years of studies into creative design practice now (Dorst 2015). They consistently demonstrate that creative design work:

- *Co-evolves*, whether this is a problem and solution space, or more appropriately a set of design arenas that do not map discretely onto problems and solutions, for example, artefacts, beneficiaries, purpose and evaluations are one possible set of concurrent design arenas that can replace crude linear problem–solution structures (Cockton 2020a).
- Successful design progressions are *concurrent*, not linear. Episodes within these progressions span more than one design arena. There are no linear phases that only focus on a single design arena at once.
- Co-evolution requires careful attention to the *balance* (mix) across different arenas for design work and their *integration*.
- By not freezing problems or requirements, designers can be *generous* and deliver beyond what sponsors and users expect or can imagine.

HCI has been slow to study design practices in detail. When detailed studies are made, the realities of creative work are exposed, such as the negotiation of what will count as a problem in usability work (Reeves 2019). Any outsider's attempt to support creative design work must be directed by insider knowledge. A researcher developing a new usability evaluation approach needs to consider findings in Reeves (2019) when framing how evaluators will integrate new approaches into their existing practice.

The secondary literature on creative design practices should be understood before committing to new primary research. As with all human-focused design, research for affective design needs to understand its users. This can be initially addressed via activities early in a research project, but it is also possible to develop understandings of intended users alongside development of novel approaches and resources. Hypothesis testing approaches from Lean UX (Gothelf with Seiden 2013) may prove to be useful here.

Research *for* design that does not include both secondary and primary research *into* design runs the same risks as any design project that ignores its potential users. Within HCI, there is no excuse at all for not allocating effort within a research project for affective design project to understanding its intended designer users.

4.5.2   Envisaging Resource Use and Checking Expectations

A research *for* design approach that is informed by 50 years of research *into* design is essential, since researchers must be able to realistically envisage how their new approaches and resources would be used in actual creative practices.

Support for designing for emotion can take a range of forms. It may be information about affective psychology, an integrative framework for analysing the likely emotional impact of an interaction design, information about options for design goals and evaluation criteria, or instrumentation for tracking emotions during evaluation. A group of resources can be co-ordinated by an approach (Woolrych et al. 2011).

Whatever the resources or approaches developed in research for design, their development must be based on a realistic understanding of how they will be used. An initial understanding can be based on the *research into design* literature (e.g., Dorst 2015). However, the main consistent results are at a very high level, for example, co-evolution, backtalk and generosity (Cockton 2020a). This limits how much research *into* design can direct research *for* design. As a result, new resources and approaches must be piloted in realistic design contexts. For example, Rutkowska and colleagues (2017) formed a multidisciplinary team across design, UX and psychology to develop a carefully designed box of cards that provided knowledge about customer loyalty, a topic within the scope of designing for emotions. This box of cards was for Pizza-Portal, a large online food delivery company. The cards were used in a design workshop with the company. They proved to be very useful prompts for the design researchers, but

> Contrary to our expectations, PizzaPortal representatives were not interested in using the cards on their own. Our tool is best suited for codesign work, and we argue that the role of design researchers is crucial for facilitating such a process. Business roles may lack the skills to use such tools unaided. PizzaPortal representatives took advantage of the knowledge presented in the cards during codesign sessions and then in the form of a written report, in which the identified mechanisms for shaping loyalty were applied to ascertain the loyalty of PizzaPortal users. The written report concluded with actionable results, and was thus a key tool for PizzaPortal representatives. (Rutkowska et al. 2017)

The cards were helpful resources for the design researchers, but the workshop approach and the report were more helpful for the company representatives. When developing support for design work, different roles will gain different value from the same resources and approaches. It is difficult to know in advance who will appreciate what and why. Novel design approaches and resources must be trialled and assessed in use (Cockton et al. 2009a, b; 2020b), regardless of the quality and status of the scientific knowledge underpinning them.

Expectations are unavoidable and will frame much research *for* design. Research *through* and *into* design are therefore essential complements to identify when, where and how expectations are mistaken.

New approaches and resources for designing for emotions must be developed and used in realistic contexts, iterating until others can use them. Worth maps evolved through supported use on four projects (e.g., George 2016), and were then used independently on several others (e.g., Camara and Calvary 2017), as well as being used effectively in teaching (Cockton 2020b). Design is an extraordinarily complex activity. Valid generalisations only hold at a high level (and normative processes tend to be invalid). It is vital to engage in concrete design settings to discover how novel approaches and resources are actually used, and what value they bring to who across a design team.

As with the first agenda item, research *for* design that does not include both secondary and primary research *into* design runs the same risks as any design project that ignores its potential users. For this second agenda item, research *through* design is also required, with novel design approaches and resources being trialled and assessed in use. Together, these two agenda items address the first two groups of issues in Sect. 4.4.

- The knowledge, experience and expertise required from a design team
- The expected working practices of design teams.

The third group of issues is addressed by the third agenda item:

- The effort required to learn and initially use new approaches and resources.

4.5.3   Collecting and Sharing Examples and Guidance

The internet has made it possible for researchers to work with a large engaged community of professionals who are interested in adopting new approaches and resources. Professional innovators can share new approaches and provide examples from themselves and others. A large set of examples can be crowdsourced.

The Persona Lifecycle: Keeping People in Mind Throughout Product Design (Pruitt and Adlin 2006) contains early examples of crowdsourced examples of completed resources. The example personas in this book were sourced from the authors' contacts in their community of practice. More recently, resource examples were collected for *Value Proposition Design: How to Create Products and Services Customers Want* (Osterwalder et al. 2014).

Dedicated online resource collections and support are more effective forms for dissemination than academic papers and theses, for example, the interactive online companion for the Value Proposition Canvas at strategyzer.com. Jones (2020) built a website for his storienteering resources (storienteer.info). One advantage of online resources is that they are accessible beyond the IxD, UX and HCI communities. Jones' resources are in extensive use in home schooling. Such use would not have occurred if he had only published his thesis (Jones 2020).

Crowdsourcing has been extended to providing evidence for evaluation. For example, resources for learning space redesign have been shared online (classroomrecipe.blogspot.com/2013/11/classroom-design-recipe.html), with examples of their use shared on Instagram (Qaed et al. 2016). This greatly extended the reach of the evaluation study, reducing the need for site visits. As with Jones, Qaed's resources have been used beyond interior design for education, including home use, and by human resource teams.

Other examples of online resource collections include two for worth mapping (Sect. 4.3 above): resourcesbyjennifergeorge.wordpress.com (George 2016) and phdgirl911.wixsite.com/arrows-and-wms (Camara and Calvary 2017). George's resources also include ones for tracking and managing balance in research through design. Support is thus available for modelling emotions in interaction contexts (worth maps) and for the progression of research through design projects.

A large community of practice needs to build around successful design practice innovations. It is important for researchers to adopt crowdsourcing and online community approaches from current professional practice. As well as hopefully demonstrating the value of systematic research approaches, feedback and examples are essential to improving resources from research for design. Family resemblances between resource examples can be reasoned about and discussed. Deliberation is a vital research resource for understanding different usage patterns.

4.6 Conclusions

Scientific knowledge on emotions, commonly from psychology, can be one basis for novel design resources and approaches. However, it is only one source of knowledge for designers and it must be combined with many other considerations. Design is a very complex activity and it cannot be focused exclusively on a single factor of specific interest to some researchers. Experimental work can greatly increase the focus on specific factors of interest, but all aspects of design need to be supported to creative professionals' standards.

It is important when looking at support design work that false assumptions and expectations are avoided. The value of some knowledge within an academic discipline, where it is recognised as the result of significant high-quality research, does not equate to its value when applied within design work. There are many choices involved as to how to translate what is often called basic, foundational or fundamental research into an effective applicable format. First wave HCI quickly encountered difficulties here, with some exploration of ways to provide structure for cognitive psychology. One outcome of research here was the highly influential task-artefact cycle with support from claims and scenarios (Carroll 1990). However, personas, which originate in industrial R&D (Pruitt and Adlin 2006), have fared far better in terms of uptake and use. High-quality science, even when translated into appropriate resources and approaches, will not automatically take precedence over resources developed within communities of practice. The latter have the advantage of being compatible with at least some ways of progressing design work, whereas resources developed outside of a community of practice must first gain entry in some way before professionals will explore their value.

Research for design must be research through design with designers. Collaboration and partnerships are essential (Camara and Calvary 2017). New resources and approaches need to be used in realistic design research projects. The creative design aspects need to be complemented by detailed tracking of activities, as in (Cockton et al. 2009a; George 2016). Records of design work form the basis for evaluating the actual impact of new innovative practices.

This chapter has been much more about design than emotions. This is because there is much more to design than emotions. New resources and approaches for designing for emotions will only inform and direct some project work. It is thus very important that new resources and approaches are actually designed for and used in realistic design contexts. Researchers must not impose their external disciplinary preferences on what are mostly effective and productive design work practices. The challenge is to imagine how knowledge from psychology or other disciplines can make an effective contribution to design work. The status of knowledge within its home discipline will carry little weight with design teams who focus on results. It is thus important to know what does carry weight with design teams and plan research with this in mind.

References

Benedek J, Milner T (2002) Measuring desirability: new methods for evaluating desirability in a usability lab setting. Presented at the Usability Professionals' Association Conference, Orlando, Florida, USA, 8 July 2002. Retrieved 20 May 2020, from: file://fileservices.ad.jyu.fi/homes/rerousi/Downloads/DesirabilityToolkit.pdf.

Boehner K, DePaula R, Dourish P, Sengers P (2007) How emotion is made and measured. Int J Hum Comput Stud 65(4):275–291

Camara FG, Calvary G (2017) Bringing worth maps a step further: a dedicated online-resource. In Bernhaupt R, Dalvi G, Joshi AK, Balkrishan D, O'Neill J, Winckler M (eds) Human-computer interaction—INTERACT 2017. Lecture notes in computer science, vol 10515. Springer, Cham. https://doi.org/10.1007/978-3-319-67687-6_8

Carroll JM (1990) Infinite detail and emulation in an ontologically minimized HCI. In: Chew JC, Whiteside J (eds) Conference on human factors in computing systems (CHI '90), ACM, pp 321–328. https://doi.org/10.1145/97243.97303

Cockburn A (2000) Writing effective use cases. Addison Wesley, Boston, MA, USA

Cockton G (2002) From doing to being: bringing emotion into interaction. Interact Comput 14(2):89–92

Cockton G (2004) Doing to be: multiple routes to affective interaction. Interact Comput 16(4):683–691

Cockton G (2006) Designing worth is worth designing. In: Mørch AI et al (eds) Proceedings of NordiCHI 2006. ACM, New York, NY, USA, pp 165–174. https://doi.org/10.1145/1182475.1182493

Cockton G (2009) When and why feelings and impressions matter in interaction design. In: Kansei 2009: Interfejs Użytkownika - Kansei w praktyce, CD Rom Proceedings, Polsko-Japońska Wyższa Szkoła Technik Komputerowych, Warszawa, Poland, 7–31. Retrieved 20 May 2020, from: https://repin.pjwstk.edu.pl:8080/xmlui/bitstream/handle/186319/126/kansei2009_Cockton.pdf?sequence=1

Cockton G (2020a) Worth-focused design, Book 1: balance, integration, and generosity. Morgan & Claypool, San Rafael, CA, USA

Cockton G (2020b) Worth-focused design, book 2: approaches, contexts, and case studies. Morgan & Claypool, San Rafael, CA, USA

Cockton G, Kirk D, Sellen A, Banks R (2009a) Evolving and augmenting worth mapping for family archives. In: Proceedings of the HCI 2009 conference on people and computers XXIII, pp 329–338, BCS. Retrieved 20 May 2020, from: https://www.scienceopen.com/document?vid=cb2684fe-e360-406f-bfb1-93bf7e64dbb5

Cockton G, Kujala S, Nurkka P, Höltta T (2009b) Supporting worth mapping with sentence completion. In: Proceedings of human-computer interaction—INTERACT 2009 II. Lecture notes in computer science, vol 5727. Springer, Berlin/Heidelberg, pp 566–581 https://doi.org/10.1007/978-3-642-03658-3_61

Doherty EP, Cockton G, Bloor C, Rizzo J, Blondina B (2002) Yes/no or maybe—further evaluation of an interface for brain-injured individuals. Interact Comput 14(4):341–358

Dorst K (2015) Frame innovation: create new thinking by design. MIT Press, Cambridge, MA, USA

Frayling C (1993) Research in art and design. Royal College Art Res Papers 1(1), pp 1–5. Retrieved 20 May 2020, from: https://researchonline.rca.ac.uk/384/3/frayling_research_in_art_and_design_1993.pdf

Gedenryd H (1998) How designers work—making sense of authentic cognitive activities. Lund University Cognitive Studies 75. Retrieved 22 April 2020, from: https://lup.lub.lu.se/search/publication/d88efa51-c2f9-4551-a259-00bd36fe8d03

George J (2016) A case study of balance and integration in worth-focused research through design [Ph.D. thesis]. Northumbria University, Newcastle, UK. Retrieved 19 May 2020, from https://nrl.northumbria.ac.uk/30326/

Gnanayutham P, Cockton G (2008) Adaptive personalisation for researcher-independent brain body interface usage. In Olsen Jr DR, Arthur RB, Hinckely K, Morris MR, Hudson S, Greenberg S (eds) CHI 2009 extended abstracts. ACM, New York, NY, USA, pp 3003–3068. https://doi.org/10.1145/1520340.1520428

Gould J, Boies SJ, Ukelson J (1997) How to design usable systems. In: Helander M, Landauer TK, Prabhu PV (eds) Handbook of human-computer interaction, 2nd edn. North Holland, Amsterdam, pp 231–254

ISO (2019) Human-centred design for interactive systems. Ergonomics of human system interaction Part 210 (ISO 9241-210). International Organisation for Standardization

Jones JC (1988) Softecnica. In: Thackera J (ed) Design after modernism: beyond the object. New York, NY, USA, Thames and Hudson, pp 216–266

Jones M (2020) Making scenarios more worthwhile: orienting to design story work, [Ph.D. thesis]. Northumbria University, Newcastle

Kalbach J (2019) Maximize business impact with JTBD. INTERACTIONS 2019 26(1): 80–83. https://doi.org/10.1145/3292021

Mitchell CT (1992) Preface. In: Jones JC (ed) Design methods, 2nd edn. Van Nostrand, New York, NY, USA, pp ix–xi

Moore H (1934) Statement for unit one. In: Read H (ed) Unit one: the modern movement in English architecture. Cassell, London, UK, pp 29–30

Osterwalder A, Pigneur Y (2010) Business model generation: a handbook for visionaries, game changers, and challengers. Wiley, Hoboken, NJ, USA

Osterwalder A, Pigneur Y, Bernarda G, Smith A, Papadakos T (2014) Value proposition design: how to create products and services customers want. Wiley, Hoboken, NJ, USA

Patton J, Economy P (2014) User story mapping: discover the whole story, build the right product. O'Reilly, Sebastopol, CA, USA

Pruitt J, Adlin T (2006) The persona lifecycle: keeping people in mind throughout product design. Morgan Kaufmann, Burlington, MA, USA

Qaed F, Briggs J, Cockton G (2016) Social media resources for participative design research. In: Bossen C, Smith RC, Kanstrup AM, McDonnell J, Teli M, Bødker K (eds) Proceedings of the 14th participatory design conference, vol 2. New York, NY, USA, pp 49–52. https://doi.org/10.1145/2948076.2948081

Reeves S (2019) How UX practitioners produce findings in usability testing. ACM Trans Comput-Human Interact 26(1):3:1–3:38. https://doi.org/10.1145/3299096

Ressler PK, Bradshaw YS, Gualtieri L, Chui KKH (2012) Communicating the experience of chronic pain and illness through blogging. J Med Internet Res 14(5):e143

Rittel H, Webber M (1973) Dilemmas in a general theory of planning. Policy Sci 4:155–169

Rogers Y, Bannon L, Button G (1994) Rethinking theoretical frameworks for HCI: report on an INTERCHI '93 workshop (Amsterdam, 24–25 Apr 1993). SIGCHI Bull 26(1):28–30. ACM, New York, NY, USA

Rutkowska J, Lamas D, Sleeswijk Visser F, Wodyk Z, Bańka O (2017) Shaping loyalty: experiences from design research practice. Interactions 24(3):60–65. https://doi.org/10.1145/3064774

Schön DA (1983) The reflective practitioner: how professionals think in action. Basic Books, New York, NY, USA

Schön DA (1992) Designing as reflective conversation with the materials of a design situation. Res Eng Design 3(1):131–147

Thieme A, McCarthy J, Johnson P, Phillips S, Wallace J, Lindley S, Ladha K, Jackson D, Nowacka D, Rafiev A, Ladha C, Nappey T, Kipling M, Wright P, Meyer TD, Olivier P (2016) Challenges for designing new technology for health and wellbeing in a complex mental healthcare context. In Proceedings of CHI '16: 2016 conference on human factors in computing systems. New York, NY, USA: ACM, pp 2136–2149. https://doi.org/10.1145/2858036.2858182

Vu P (2013) A worth-centered development approach to information management system design [Masters Thesis]. Aalto University Press, Helsinki. Retrieved 20 Apr 20 2020, from: https://www.soberit.hut.fi/T-121/shared/thesis/di-Phuong-Vu.pdf

Whiteside J, Bennett J, Holtzblatt K (1988) Usability engineering: our experience and evolution. In: Helander M (ed) Handbook of human-computer interaction, 1st edn. North Holland, Amsterdam, pp 791–817

Woolrych A, Hornbæk K, Frøkjær E, Cockton G (2011) Ingredients and meals rather than recipes: A proposal for research that does not treat usability evaluation methods as indivisible wholes. Int J HCI 27(10):940–970

Chapter 5
An Innovative Humour Design Concept for Depression

Huili Wang and Xueyan Li

Abstract This chapter focuses on the design ideas of collaboration between the fields of design science and neuroscience to seek approaches to help people with depressive mood. Cutting-edge neuroscience research, involving sophisticated brain imaging techniques, can offer more and deeper insights into the cognitive processes of human being. These techniques explore different features in brain signals and functional brain regions to measure the biological factor changes related to the processing characteristics of depressive mood. Under the guidance of life-based design (LBD) and individual knowledge system (IKS) theories, resorting to mobile phones as the main media connecting potential users and future technical artefacts, the model proposed in this chapter was integrated with the emerging EEG-based brain–computer interfaces to provide a 'four-in-one' system: real-time recording, instant feedback, effective intervention and constant reinforcement to optimise the way to bridge scientific studies and real lives and investigate solutions of mood disturbance.

5.1 Introduction

Depressive moods exert negative effects on our life and work. Design researchers have sought to use current technologies to develop innovative and feasible design concepts to produce practical technical artefacts that improve the quality of life for people with depressive mood. This chapter examines recent collaboration between the fields of design science and neuroscience to seek approaches to help people with depressive mood by utilizing humour to positively and actively influence their physical and mental health and enhance their social interactions. Studies of the cognitive

H. Wang (✉) · X. Li
School of Foreign Languages, Dalian University of Technology, Dalian, China
e-mail: huiliw@dlut.edu.cn

X. Li
e-mail: lixueyan@dlut.edu.cn

© Springer Nature Switzerland AG 2020
R. Rousi et al. (eds.), *Emotions in Technology Design: From Experience to Ethics*, Human–Computer Interaction Series,
https://doi.org/10.1007/978-3-030-53483-7_5

processing mechanisms in humour appreciation provide knowledge systems, evaluation methods and intervention tools to bridge the gap between neuroscience and design science research. The objective is to facilitate creative design processes that help transfer the positive influences of humour in relation to technical artefacts by helping people regulate and manage their emotional states.

We propose an innovative humour design concept embodied in the 6Rs model on the basis of life-based design (LBD) theory. Given that mobile phones are the main media connecting potential users and future technical artefacts due to their advanced functions, such as instant communication, portability, video in real time, privacy protecting, and so on, the model, via mobile phone, is integrated with emerging electroencephalograph (EEG)-based brain–computer interfaces (BCI) and individual knowledge system (IKS) to provide a 'four-in-one' system: real-time recording, instant feedback, effective intervention and constant reinforcement. The model is designed to lay more objective and reliable foundations for the design and implementation of future technical artefacts.

5.2 Design Concept in Design Science

5.2.1 Life-Based Design and Form-of-Life Analysis

Technologies, which combine human actions with technical artefacts or systems, are ultimately designed to enhance people's quality of life (Saariluoma and Leikas 2010). Design practice should relate to people's everyday experiences and their relevant social, cultural and mental aspects to the contexts of technical artefacts (Saariluoma et al. 2016). Therefore, LBD, as a basis for the creation of design ideas and for concept design, defining the particular problems being solved and the role of technology being related to, involves a multi-dimensional design theory that highlights the significance of comprehensively understanding people's lives, including forms of life, values and circumstances to ensure that the artefacts focus on better satisfying people's needs (Leikas 2009; Rousi et al. 2011). On the basis of ontological thinking, the essential element of LBD is to discover and identify information about forms of life (life circumstances), values and people's actions collected at the initial stage of the design process, and then efficiently integrate this information into the design to improve technology-supported actions present in people's everyday lives (Saariluoma and Leikas 2010).

Thus, four major design issues in LBD are: form-of-life analysis, concept design and requirement generation, fit-for-life analysis and innovation design (Rousi et al. 2011). Form-of-life analysis is important for LBD since it exposes relevant differences between different life settings, which can provide a precise yet elastic enough concept to define the target (Saariluoma et al. 2016), and the subsequent three issues will be set up on the basis of form-of-life analysis.

Diversified aspects associated with people and their living states constitute human life sciences, presented by the form of life encompassing three essential factors – biological, psychological and socio-cultural factors (see Figs. 5.1 and 5.2). The *biological* factor is perhaps the most important due to its investigations in ergonomics (Akerstedt 1985), physiology (Mykles et al. 2010), anatomy (Pelphrey et al. 2005), neuroscience (Giese and Poggio 2003), ecology (Berkes and Turner 2006), medicine and health care (Kaslow et al. 2007). The *psychological* factor concerns perceptions, emotions and behaviours embodied in social psychology, philosophy and education. *Socio-cultural* factors relate to sociology and culture including gerontology, anthropology, literature, semiotics, film research, art, ethnography and linguistics (see Fig. 5.2). These three factors are indispensable to understanding the basic structures of the form of life. Design science that seeks to serve human beings should be based on holistic and objective investigations into the form of life to ensure the technical artefact satisfies users' needs. Therefore, form-of-life analyses should precede every experimental process, concept design or technology innovation to propose more practical, valid and concrete technical artefact designs to enhance peoples' well-being.

Given the speed at which science and technology develop, design theories must be kept up to date in order to enrich the methodological processes. Past investigations into the form of life were mostly achieved by surveys and interviews. To some extent, these methods cannot objectively reflect reality due to the lag in recording. Furthermore, those closely involved cannot see clearly, so more objective methods of

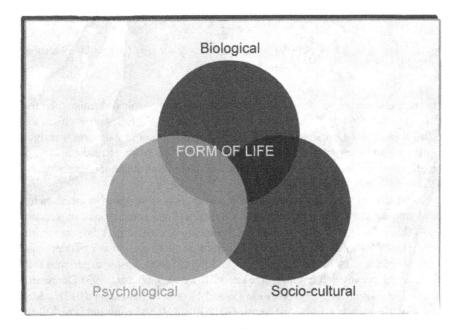

Fig. 5.1 Essential factors of form of life (Leikas 2009, p. 88)

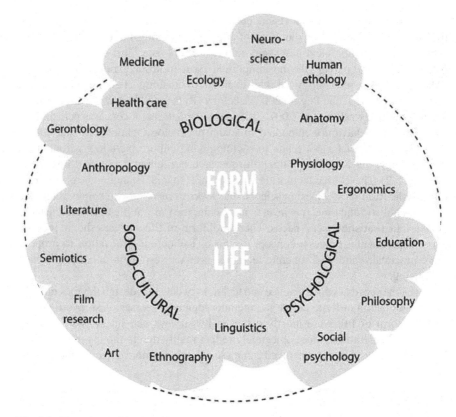

Fig. 5.2 The palette of human life sciences connected to research on forms of life (Saariluoma et al. 2016, p. 192)

observing and analysing forms of life are needed to help people determine their true states of mind. Cutting-edge neuroscience research, which relies on sophisticated brain imaging techniques, can fill this gap by offering more (and deeper) insights into the cognitive processes. These techniques explore different features in brain signals and functional brain regions to measure the biological factor changes related to the brain in the LBD analysis.

Negative moods, concerning depression, anxiety, stress, poor self-respect and low self-confidence, can affect an individual's judgement and perception of objects and events (Laceulle et al. 2015), which can cause them to misinterpret the surrounding world. Up to 25% of the population will experience at least one episode of depression in their lifetime (Kessler et al. 2001); up to 30% of young people experience mild subclinical depressive symptoms by the age of 18 (Lewinsohn et al. 1993). Depression affects 350 million people worldwide (World Health Organization 2012), and is expected to be the second-greatest cause of ill health in 2020 (Leahy and Dowd 2002). The term depression refers not only to the clinical disease, which often prevents people from living a normal life, but also to a feeling of unhappiness or sadness that deprives people of their hope for the future. Non-depressed individuals also experience mood swings and may have 'blue' hours or days that are similar in a number of ways to clinical depression (Beck and Alford 2009), which can seriously impact their mental health. This chapter thus seeks ways to alleviate the problem of depressive mood.

5.2.2 LBD Analysis on Depressive Mood

We conducted detailed LBD analyses on people who suffer from depressive mood in terms of biological, psychological and social–cultural factors and propose a holistic approach that precedes the proposal of the following humour design concept.

The first factor is the biological factor. Though epidemiologic studies show that roughly 40–50% of the risk of depression is genetic (Detera-Wadleigh et al. 1998), non-genetic factors during brain development are also involved (Akiskol 2017), revealing the close correlations between depressive mood and the brain. Drevets (2001) demonstrated that the changes in some brain areas were highly correlated with depression, such as in the 'prefrontal and cingulate cortex, hippocampus, striatum, amygdala and thalamus'; anatomic studies of the brains of people with depressive mood obtained at autopsy also reported abnormalities in many of these brain regions (Zhu et al. 1999). In addition, Nestler et al. (2002) indicated that particular parts of the brain were connected to the mediation of different aspects of depression, and different brain areas operated in a series of highly interactive parallel circuits to formulate the neural circuitry involved in depression.

The second type of factor is psychological. According to Beck (2011), people with depression often formulate 'if–then' statements (e.g., 'If I don't do as well as others, then I'm a failure' or 'If I trust others, and then I will get hurt') and have difficulties expressing the nature of reality without considerable outside help because of their 'deep cognitive structure'. They are also often unwilling or unable to examine the evidence against their validity, since they consciously think about them as 'just the way things are' (Leahy and Dowd 2002). The chief psychological complaints of depressive people include 'I have nothing to look forward to', 'afraid to be alone', 'no interest', 'can't remember anything', 'get discouraged and hurt', 'black moods and blind rages', 'I am doing such stupid things', 'I am all mixed up', 'very unhappy at times' and 'brooded around the house' (Beck and Alford 2009). Furthermore, Rissanen (ND) indicated that several elements become greatly distorted in the course of the psychological cognition of people with depressive moods, such as negative automatic thoughts, beliefs, rules for living and core beliefs in daily life (Beck and Alford 2009). Depressed people are also generally more sensitive to failure than non-depressed people. They react with significantly greater pessimism and a lower level of aspiration, and tend to distort their experiences in an idiosyncratic way: they misinterpret specific events in terms of failure, deprivation or rejection, and are inclined to make negative predictions about their future (Beck and Alford 2009).

The third factor is social–cultural, with deeply-rooted distorted mentality. Unhappy personal experiences, severe stress from studying and work, negative influences from violence, neglect or poverty may make people vulnerable to depressive moods. Their ordinary states are feelings of sadness or a marked loss of interest in (or pleasure from) social activities, with a sense of worthlessness or inappropriate guilt. Research on day-to-day social interactions has revealed that, compared to non-depressed participants, depressed participants found their interactions to be less enjoyable and less intimate, and they felt less influence over their interactions (Nezlek et al. 2000). Depressed people usually had more intense responses to negative interactions with people. Some studies suggest that depression may sensitise people to everyday experiences of social rejection or social acceptance (Steger and Kashdan 2009). With a decreasing reliance on interpersonal relationships outside the family, Forgas et al. (1984) collected evidence that people with depression were critical of their surrounding happenings, and showed an excessive tendency for mood fluctuation. Meanwhile, they engaged in more frequent online social interactions, which generated negative effects associated with internet use (Caplan 2003).

In summary, in terms of biological factors, to find an effective way to conquer depressive mood, an innovative design concept that entails proper interventions against the current technological background is needed to mediate the neural circuitry to positively stimulate activities in relevant brain areas. In light of psychological factors, we need to effectively convert depressed people's defective and partial thoughts into positive ideas by using an intervention tool with positive stimuli to change their ways of thinking and shift the angles from which they view the problems. From the angle of social–cultural factors, the habitual approach to communication in interpersonal relationships can be improved to increase their opportunities to socialise with people outside their limited social networks.

Humour, as an effective 'medicine' and a frequently used intervention method, can be used to mediate the biological functions of relevant brain areas of people with depressive mood, influence their ways of thinking and engage them in more social interactions, helping them to build up their confidence in communicating with other people.

5.2.3 LBD and Individual Knowledge System

IKS is an artificial intelligence (AI) oriented generalised information and knowledge system owned by one person that includes all memories of the experience, information and narrow sense knowledge supported by the system including the body, especially the brain. Most importantly, it is an evolving, integrated self-organisation system. As shown in Fig. 5.3, in the IKS framework, the objective world is the object of cognition, and group knowledge (including the family culture, national culture, social knowledge, etc.) is its environment. IKS is supported by the body's intelligence hardware, which consists of three layers: the central brain neurobiology system of memory and computing; neuron sensory memory and neural network; and

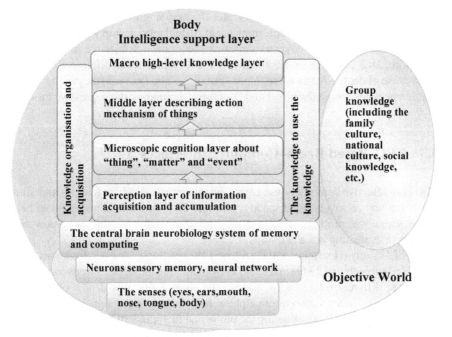

Fig. 5.3 The IKS framework (Wang et al. 2020)

the senses (eyes, ears, mouth, nose, tongue, skin and body). It is mainly made up of six epistemic parts, including a perception layer of information acquisition and accumulation; a microscopic cognition layer about 'thing', 'matter' and 'event'; a middle layer describing the action mechanism of things; a macro high-level knowledge layer; knowledge organisation and knowledge of learning; and knowledge of how to use the knowledge (Wang et al. 2020).

The computation model (Basic Knowledge Element Model) of the six-space pattern of IKS includes data space, metadata space, instance model space, formal model space and operator space. The six-space pattern P^{6S} as a generalised representation of knowledge in four levels for IKS (see Fig. 5.3) is a computation model that individuals can use to analyse things. Each space is a set characterised by the multi-level elements corresponding to knowledge, and the different mapping relationships to various attributes associated with the set. Therefore, it can be used as a fundamental management frame of AI—a basis on which to manage and analyse big data (Wang et al. under review). Due to the time lag and subjective bias in traditional interviews or reports used to analyse psychological and socio-cultural factors, P^{6S} will integrate comprehensive data related to the six spaces into a more objective reflection of the form of life of the target population by big data and AI.

Four merits relevant to LBD in terms of IKS analysis and BCI are as follows. First, the IKS framework described in Fig. 5.3 shows that the three factors of LBD analysis are biological, psychological and social–cultural factors. Second, IKS is AI based. Third, the IKS framework illustrates the mapping of the objective world in the subjective knowledge domain, emphasizing a dual relationship with the objective world. Fourth, the IKS pattern can enable more accurate and objective computations. Therefore, IKS can help more objectively identify individual differences and the exact cause of depressive mood so that we can make adaptations to the psychological interference, enhancing the effect and hopefully eliminating the depressive mood.

5.3 EEG-Based Brain–Computer Interfaces

EEG is an effective commonly-used brain-imaging method, which features high time resolution and relatively good space resolution. EEG-based BCI research can help relieve emotional problems and even disordered mental states due to its powerful features in evaluating brain activities.

An EEG study (Henriques and Davidson 1991) explored the brain functions of 15 depressed subjects and 13 normal subjects by extracting the relevant frequency bands with three different reference montages. The study found that depressed subjects had less left-sided activation (more alpha activity) than normal subjects, which was recognised as a deficit in evaluating the biological factor in LBD analysis and can be used as an indicator in designing artefacts related to people with depressive mood. Another EEG study revealed that people with relatively stable left frontal activities reported greater positive emotions to positive films, while people with relatively stable right frontal activities reported greater negative emotions to negative films (Wheeler et al. 1993), suggesting that the left hemisphere is more responsible for the formation of positive emotional states, while the right hemisphere is more responsible for the formation of negative emotional states. This notion of 'positive emotion = left frontal' versus 'negative emotion = right frontal' asymmetry has been referred to as the 'affective valence hypothesis' of frontal EEG asymmetry (Harmon-Jones et al. 2009), which has great reference values in evaluations of emotion-related artefact designs. A recent study showed that the activity level of the right brain of depressive people needs to be improved, so that the enhanced synchronisation performance in the right hemisphere—correlated with some abnormal areas in the right brain, such as right frontal gyrus, triangular part of the right frontal gyrus, and orbital part of the right inferior frontal gyrus—will help to overcome the functional barriers in these brain areas (Li et al. 2018).

5.4 The Merger of LBD, EEG-Based BCI and IKS

If cognitive science should, therefore, study the mind not in isolation but in interaction with the physical world, then it is a natural second step to ask how to design artefacts that can best facilitate this interaction (Parasuraman 2008). A practical and effective artefact design that promotes interactions between cognitive study and realities also requires guidance from theory. It should also be combined with the updated technological contexts, and should precisely locate the burning problems that have serious negative impacts on people's living states. LBD and IKS theories, which involve the comprehensive analysis of LBD on target people, are indispensable to a design concept in a holistic artefact design. Brain-imaging techniques like EEG, which can objectively record and evaluate real-time LBD states in terms of brain responses, can help develop an effective artefact design. The interface between the brain and the computer is a prerequisite to this process; the construction of the IKS provides an objective AI-based methodology to identify the exact causes of the depressive mood based on the knowledge system of the depressed individual. We propose a humour design concept that merges EEG-based BCI, LBD and IKS with humour research findings to create an innovative, optimised way to bridge the scientific studies and real lives to provide solutions to the social issue of mood disturbance.

The EEG-based BCI, LBD and IKS in this humour design concept are theoretically fundamental and methodologically complementary to each other. A practical BCI artefact must be designed on the basis of life and take into account potential users' biological, psychological and social–cultural factors. Investigations into the social–cultural factor can mostly be reached by subjective self-reports or surveys; EEG has more objective and reliable advantages in observing the mental states of people with mood problems; IKS helps objectively recognise the cause of the depressive mood and excels in AI-based calculations in terms of knowledge elements of the depressed individual, making up for the disadvantages in subjective recordings of brain responses.

5.5 Mobile Phones and Mental Health

Mere investigations into LBD analysis and the integration of EEG-based BCI and AI-based IKS cannot guarantee a successful design outcome. Well-grounded methods or tools are needed to further connect users with the technical artefacts and facilitate communications between them. Information communication technology (ICT) is one of the innovations having the greatest effects on people's lives in recent decades. The majority of people, especially youth, depend on ICT to accomplish most of their daily activities. ICT is no longer a tool to assist them in their lives, but a popular way of life.

Mobile phones are currently the most frequently used ICT devices. Recent research shows that even people with severe psychiatric disabilities and functional impairment, as well as many unsheltered homeless individuals, own and use mobile phones (Ben-Zeev et al. 2013). In addition to commercial purposes, mobile phones can also be utilised to help in health care. They can be carried by the person and facilitate bidirectional communication and on-demand access to resources (Proudfoot 2013). Average users check their phones as often as 150 times a day (Meeker and Wu 2013), which reflects the great extent to which smartphone applications (apps) can generate, reward and help to maintain strong habits involving their use

(Oulasvirta et al. 2012). Apps can be used to implement cognitive and behavioural interventions to improve users' physical health (Free et al. 2013) and mental health as well. Thus, mHealth (mobile health care), a rapidly growing area that relies heavily on mobile applications and handheld devices, represents a new frontier for delivering mental health treatment (Kazdin and Blase 2011). Mobile phone health technology has great potential to facilitate the management of emotional health through its ability to deliver flexible, user-oriented intervention and self-management tools since young people tend to choose non-professional or self-managed strategies to deal with their mental health issues (Rickard et al. 2016).

With the rapid upgrading of mobile phone technology and the development of apps, there are several mental health apps available. The demand for mHealth apps is strong, as evidenced by a recent public survey in which 76% of 525 respondents reported that they would be interested in using their mobile phone for self-management and self-monitoring of mental health if the services were free (Bakker et al. 2016). These apps can remind users of their daily schedules and perform targeted interventions in their mental disturbance to enhance positive moods. Ben-Zeev et al. (2013) found that users and providers can employ smartphone health apps for diagnostics, behavioural prompts, reminders, continuous illness monitoring and self-management programs that extend well beyond the boundaries of a physical clinic.

However, many apps lack experimental validation and efficacy. One of the reasons is that the developers and the users of the apps have been pushed back at the control of the App Store's policies of paying for the app trials and usage so that a potential promising market waiting to be developed is formed. Thus, we need to propose a practical and valid concept design, which will produce prototypes and then technical artefacts to help people with negative moods by providing them with an easy-to-reach and effective way to cope with the problem based on LBD with cognitive interventions via mobile phones.

5.6 A Concept Design of Humour Implantation in mHealth

The LBD concept (Saariluoma et al. 2016) elaborated on the process of forming a design concept and proposed some sequential steps to follow. The first step is to define the problem to be solved. Then comprehensive investigations are to be conducted by connecting to the LBD and IKS analysis of the target population, including their conditions and individual attributes in terms of biological, socio-cultural and psychological factors. A concept design will then be proposed based on the previous findings, and a detailed description of how the new design concept can improve human actions will be illustrated. The third step relates to the design process, in which the design plan is divided into sub-problems; these problems are to be solved based on the literature on the current technical contexts. Then, the new knowledge or innovative design will be worked out, depending crucially on the needs and specific situations of the target users. The relevant new technologies will be generated, and how these technologies can be linked to the problems will be illustrated in great detail, including a more concrete implementation plan and specific illustrations for each step forward, such as the relevant technical staff involved and the equipment or devices employed. Finally, the proposed innovative design will be refined based on feedback from potential users' experience or evaluations by modifying the previous concept design. After the round of this cycle of concept design –> innovative design –> user experience –> refined concept design –> refined innovative design, the final innovative design concept will eventually be developed (see Fig. 5.4).

In this humour design concept that seeks to overcome negative moods, humour is adopted as the principal solution. Humour concerns a high-level cognitive process that comprises emotional, behavioural, physical, psychological and social aspects. Freud (1960) described humour as a specific defence mechanism, through which positive emotions can overcome undesirable negative emotions in stressful situations. Beck (1991) emphasised that when dealing with people who suffer from depression, we must never lose sight of the gravity of their loss: the constriction of the capacity to feel pleasure, affection, gaiety and amusement. Laughter or mirth, caused by humour, makes people feel more relaxed, cheerful, less depressed and less stressed. It is one of many largely hardwired behaviour patterns that humans use to communicate a wide range of positive emotions and its prominent function is related to joy. The peculiar sounds of laughter have a direct effect on the listener, including positive emotional arousal that mirrors the individual's emotional state by activating certain specialised brain circuits (Gervais and Wilson 2005). Research on the neural mechanism of humour processing has found that exposure to a four-minute humorous film led to a significant reduction in reported feelings of anxiety relative to baseline (Moran 1996). In addition, humour produces positive short-term emotional changes that are at least comparable to (if not superior to) the effects of vigorous physical exercise (Martin 2010).

Different humour styles have different influences on positive moods. Affiliative humour, used to create more harmonious bonding with other people, is prosocial and positive to enhance social interactions; self-enhancing humour is used to boost confidence in difficult situations. The humour-related positive emotion of mirth also helps achieve the basic tasks required for various relationships. Mirth accompanied by humour replaces the negative feelings that would otherwise occur, enabling the

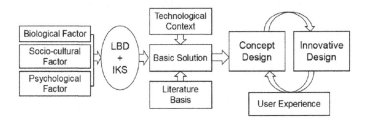

Fig. 5.4 The flow of concept design and innovative design

person to think more broadly and flexibly and to engage in creative problem-solving (Fredrickson and Joiner 2002).

Several studies have found that humour functions as an emotion regulation mechanism, reducing negative emotions and enhancing positive moods. Cann and Calhoun (2001) compared the effects of exposure to humorous versus a neutral videotape after participants watched a film that relayed negative feelings. The humorous video produced lower ratings of depressive mood and anger and higher positive moods compared to the neutral video. The researchers concluded that the elevated positive emotions associated with humour may counteract feelings of depressive mood and anger, and the effects of humour may be more cognitively mediated: humour preceding the stressor might work as a cognitive prime, changing the way subsequent events are interpreted and thereby reducing anxiety (Martin 2010). On the whole, humour helps to cope with stress, smooth social interactions, increase creativity and enhance life satisfaction and well-being (Shen et al. 2015).

Based on the theory of humour, the incongruity theory, the key of the three-stage model (incongruity detection, incongruity resolution and mirth) is to complete the appreciation of the humorous material by finding a new path to resolve the task. This predominant theory in this field illuminates the perceptual–cognitive processes of humour appreciation, which makes us view people, situations and events from the perspective of two or more incongruous (and seemingly incompatible) perspectives at the same time and solve the problem by understanding the new relationships among these originally incongruous things. Thus, understanding humorous material will help people who suffer from depression transform their old way of thinking and feel relaxed by finally laughing as well. The activation of certain brain regions related to humour appreciation will stimulate the neurons of brain regions associated with depression, which will help revert the depressive state to the previous healthy state.

People with psychological problems are not very motivated to change, because their problem behaviours result in too many rewards for themselves; alternatively, they might be afraid of the unknown that the change will bring about (van Bilsen 2013). Whereas, when people tell jokes, they seem to be able to control and manage the situation even if they need to face the change so that humour interventions can be used as both clinical and non-clinical treatment and prevention to enhance people's positive moods.

Mobile phones are fully integrated into modern life, which can facilitate the user's/provider's communication and the delivery of time-sensitive health information. More researchers from different health disciplines are interested in developing evidence-based mHealth interventions for a range of physical and mental health conditions (Heron and Smyth 2010). By integrating today's technical context, such as the EEG-based BCI and mHealth, with the positive functions of humour on negative mood, a humour design concept on the basis of LBD and IKS can be implanted into mHealth.

5.7 A Humour Design Concept: 6Rs. Model

The design concept is a meta description of purposes and scopes as well as the principles of forms and functions. It adds prescriptive knowledge to a certain system and constitutes guidance towards a nascent design (Adam et al. 2017). The goals of this humour design concept are to enhance the level of positive emotion among people and to improve their physical and mental health. One of its main components is to record real-time biosignals—such as EEG signals, heart rate and eye-tracking information—to transfer the analysed results, combined with subjective investigations, into countermeasures against mood problems. The 6Rs. model mainly leverages neuroscience as the basis of designing the conceptualisation of the emotion-related technical artefact through using brain-imaging methods as both built-in functions and evaluation methods.

With the assistance of more sophisticated and advanced technological methods, real-time biological signs such as brain signal, heart rate, blood pressure, skin conductance and eye-tracking can be recorded simultaneously in real time. The exponential development of brain-imaging techniques, computers, sensors, robotics and AI, in power and performance as well as a reduction in energy and cost, combined with the ubiquity of communication technologies like the internet, has triggered a meteoric rise in smart connected devices such as wearable monitors, smart watches and smartphones (Hird et al. 2016). These developments have armed researchers with the most effective tools and media to help collect biological data and provide feedback. EEG has been developed and upgraded to be more portable and easier to use; for example, it can be hidden inside a hat. By locating regions of interest in the brain, the portable EEG will effectively collect signals related to brain activities to provide data to help supervise the user's emotional states.

Eye movements pervade visual behaviour and are viewed as a significant aspect of human behaviour and a window into the perceptual and cognitive processes underlying behavioural performance (McCarley and Kramer 2009). The visual scene is typically inspected with a series of discrete fixations separated by rapid saccadic eye movements; information is collected only during the fixations. Visual input is suppressed during the movements themselves, and the observed eye movements are controlled by a broad network of cortical and subcortical brain regions, including the parietal cortex, the prefrontal cortex and the superior colliculus (McCarley and Kramer 2009). Sophisticated eye-tracking systems are now abundant, inexpensive, easy to use and often even portable.

In addition to portable EEG and eye-tracking systems, other sensors can be used to collect different biosignals. For example, JWatcher (https://www.jwatcher.ucla.edu/), an event-recording software, can be interfaced with video or automated sensor data to provide synchronised records of behaviour and physiology that are essential to link overt actions with underlying mechanisms (Parasuraman and Rizzo 2009). Similar wearable devices can be put into clothing, such as heart sensors in bras. Multiple devices are gradually being developed and integrated into a single device that is portable or wearable. The increase in the availability of sensors to measure common vital signs and behaviour in unobtrusive forms worn on multiple body locations has led to a surge in commercial monitoring applications for wellness and health (Hird et al. 2016). Users of these devices and sensors can record their real natural behavioural states anytime, anywhere. However, the strengths and weaknesses of different tracking systems, interfaces and synchronisation between different data streams—and the pros and cons of different sensors (e.g., optical, audio, electromagnetic, mechanical), sensor calibration and drift, and data acquisition and analysis software—are important topics for further research (Parasuraman and Rizzo 2009).

Most current emotion-related apps only include subjective emotional evaluation; they generally do not offer systematic suggestions or solutions to the problem, which to some degree increases depressive people's feelings of helplessness, frustration and hopelessness. The humour design concept proposed here includes more objective, accurate and practical evaluation methods and intervention strategies by employing multi-measurement tools to provide more comprehensive solutions to the problems. The principal design intention is to shift more attention and resources to preventive care and early intervention to help people detect problems earlier and overcome the damage brought about by negative moods. It provides a real-time feedback and mobile-supervising system that is responsive to individuals' physical and mental signals. If abnormal signs are detected, mobile interventions will be triggered, and coping strategies will be provided to ease the emotional fluctuation.

This humour design concept consists of the following steps: data input, data analysis, data results output, psychological consultant's solution and humour intervention (see Fig. 5.5). The design concept features a mood kit, which includes optional subjective self-reports about the user's mood state. Via the mood kit, the subjective evaluations of users' mood states are transferred into data and collected by the analysis software (see ① and ③ from Fig. 5.5). The mHealth app will be connected to the video function, and the facial expression recognition data will be transferred to the data analysis software after the selfie (see ② and ④). Then, using external portable equipment, such as EEG (dry/wireless wearable electrodes), ECG (biosensors), eye-tracking systems, etc. the relevant real-time data on reading the specified material

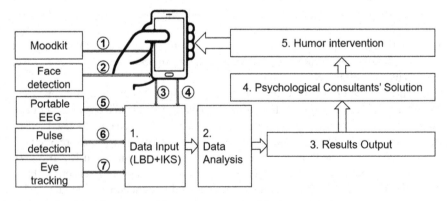

Fig. 5.5 The flow of data collection and data analysis of the humour design concept

programmed is transmitted to the data analysis software (see ⑤, ⑥ and ⑦). After the comprehensive data collection on biological, psychological and social–cultural factors, the researchers will conduct LBD and IKS data analysis and evaluate the mood states compared with the individual's normal-standard mood state values set in advance based on the individual's big data. The data analysis results will then be passed on to the researchers and psychological consultants for further integration and analysis. The consultant will engage in comprehensive evaluations to suggest solutions to counter the negative mood via a humour intervention. Each of the 6 'Rs' associated with this humour design concept is discussed in turn below.

(1) *Record real-time biometric signals and physical signs*

External portable equipment, such as EEG, ECG, eye-tracking systems and facial expression recognition systems (embedded in the app), will be employed simultaneously or selectively (according to the situation) to collect real-time biometric signals and physical signs. The sensors in the devices operate wirelessly; most applications have built-in Bluetooth technology to receive and transmit data via the internet or Wi-Fi to a local, dedicated web or cloud server for storage (Hird et al. 2016). For example, emotional changes will be detected by the EEG analysis of the slow-frequency waves (delta and theta band) and left or right frontal areas of the brain (Davidson 1993). Time windows of 400 and 600 ms can detect the integration of language processing, and Broca's area and Wernicke's area are significant regions of interest for detecting language abnormalities. If users have a depressive inclination, the hippocampus and hypothalamic–pituitary–adrenal axis, which are related to the brain's reward pathway, should be monitored and measured.

The actual EEG data extraction and analysis are far more complicated, requiring a team of professional researchers to set more precise standards on time, space and frequency dimension targeting different cognitive tasks to determine the user's emotional state. Since the normal standards of emotional states have distinct individual differences, depending on the analysis of the user's personal big data, the normal standards can then be set based on specific factors such as gender, age, personality and handedness. The development of advanced techniques for high-dimension analytics, ranging from inference techniques to machine learning and AI methodologies, provides powerful tools to identify patterns in the data (Hird et al. 2016). IKS will also generate more comprehensive data sources to guarantee more objective analysis of brain states and psychological situations.

(2) *Retrieve the mood state by subjective self-report*

Though self-reporting of mood states has its drawbacks, which may give rise to invalid evaluations due to delayed recall and bias, this most frequently used mood state measurement has great reference values to reflect biological, psychological and social–cultural factors of LBD to some degree. Thus, in this design concept, subjective self-reporting is used as an auxiliary to the measurement of real-time biometric signals to supplement the absence of the brain-imaging tool or sensors. The mood kit adopted in this design concept will be a multi-dimensional mood scale. The relevant data will be compared with a pre-set standard value (extracted from big data analysis), and the results will be combined with those from the biosignals to assess users' mood states.

(3) ***Repeat measurements over time***

In order to collect relatively stable and reliable data, the previous measurements should be repeated over time to collect the necessary dynamic information on fluctuating moods to determine individuals' mood state baselines. Furthermore, repeated long- and short-term measurements allow researchers to monitor how people's mental states change over time. The specific times and places of data collection depend on several conditions, including the specific situation (e.g. Are the times and places suitable for the usage of the relevant equipment?), the evaluation of results and users' willingness, which is complicated to determine and requires further research to set specific standards for different individuals.

(4) ***Regulate the individual indexes of depressive mood***

Resorting to IKS, individual depressive mood indexes can be regulated. Humour can help moderate moods in a positive way for a certain period of time, but it cannot completely eliminate depressive mood. Identifying the underlying causes will be necessary to solve the problem. IKS can achieve this by computing the results of an individual's knowledge system based on his or her knowledge space, including data space, metadata space, knowledge element space, instance model space, operator space and formal model space.

Figure 5.6 illustrates the computation process. First, we need to obtain the basic knowledge element set that describes those kinds of objects: specifically, the attributes set A_e and the relation set R_e of every kind of object can be

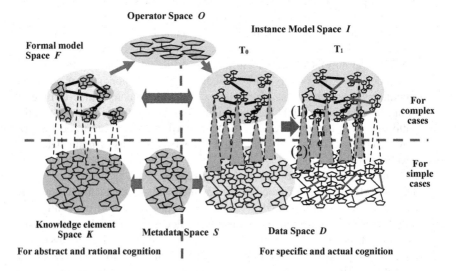

Fig. 5.6 Deductive process of the computation mode (Wang et al., under review)

acquired through theoretical and empirical knowledge, so knowledge element space K may be constructed. Then, the system of objects can be enumerated and described by data space D and metadata space S based on instantiation of the knowledge element. For simple or common cases, human consciousness and behaviours can be explained by the three spaces of K, D and S. Additionally, the basic knowledge element set in the data, metadata and knowledge element spaces coincide with the biological, psychological and social–cultural factors. Thus, computing the three spaces allows us to identify the IKS of the target person with depression and, most importantly, identify the root cause of the depression, which facilitates any subsequent psychological intervention (Wang et al. 2020).

(5) ***Repair depressive mood by humour intervention***
Fostering humour by mobile phone under the supervision and instruction of psychological consultants is the principal intervention method. The solutions will be based on McGhee's humour intervention program and the Seven Humour Habits Program. The essence of fostering humour is to train people to appreciate and produce humour in order to improve their mood. Ruch and McGhee (2014) developed a humour intervention program that emphasises strengthening key humour habits and skills in a hierarchical order (establish a playful attitude, laugh more often and heartily, create verbal humour, look for humour in everyday life, laugh at yourself and find humour in the midst of stress). Their program is a standardised training program that can be completed individually through a manual or guided by an instructor in a group (Proctor 2017) via mobile phone apps. The program has demonstrated its effectiveness at increasing positive emotions and subjective well-being in both healthy adults (Crawford and Caltabiano 2011) and clinically depressed adults (Falkenberg et al. 2011). Moreover, it decreases seriousness and negative moods (Falkenberg 2010), depression (Crawford and Caltabiano 2011) and anxiety (Proctor 2017).

As mentioned above, EEG-based BCI in this design concept is not only a method of collecting objective data. It is also an effective tool to help analyse the data in comparison with the stored information to inform the corresponding intervention. For example, prior to the intervention, the EEG data analysis will identify which brain areas are relevant to the problematic emotions; during the course of the humour intervention, the EEG data collection will be monitored to detect changes in the time, space and frequency domains. After the intervention, through the data analysis, the findings related to the cognitive processes involved in humour appreciation will be designed to be applied to evaluate whether the intervention exerts the desired influences on the individual's emotional state. For humour processing among normal people, a bilateral activation during the cognitive stage in the left hemisphere network shows stronger activation in humour detection and comprehension, especially when related to verbal jokes, focusing on the speech-related regions of the brain. The mesolimbic reward regions, with the highlights of the amygdala, hippocampus and insular cortex, are the emphasis areas to monitor for the affective component of humour appreciation. In terms of the frequency domain, for example, if the stimuli are verbal jokes, beta, one of the frequency bands, should be paid more attention due to its close involvement in language processing. These brain networks and particular brain regions are the focusing areas in collecting data; at the same time, they are the reference areas used to evaluate the abnormality analysis and intervention effects. The actual situations will be more complex.

In addition to the human intervention of humour, the computational humour intervention is also included in this design concept. By running the computational humour models, the app in the design concept will also serve the functions of joke understander and joke generator to immediately respond to the users' language input in humorous ways. This will not only help foster the users' sense of humour; it will also increase the communication opportunities in the virtual reality environment.

The human intervention and computational intervention both depend on the selection of the jokes with varying types and degrees of humour. IKS computation can help the psychological consultants choose appropriate jokes based on the results of the individual differences.

(6) ***Remind users regularly to incorporate humour into their lives***

Incorporating humour into everyday life not only helps people improve their social interaction skills and more importantly, it decreases negative moods by applying the humour habits to daily life. The current design concept is to shift more attention and resources to preventive care and early intervention to help people detect problems earlier and overcome the damage brought about by negative moods by providing a real-time feedback and mobile supervising system that is responsive to each individual's physical and mental signals. Once the abnormal signs are detected, mobile interventions will be triggered and coping strategies will be provided to help ease the emotional fluctuations. However, people's attitudes towards depressive mood are very negative. Due to fears of prejudice and discrimination, even when services are available, three out of four people suffering from depression avoid or delay treatment, resulting in a failure to receive the help they need (World Health Organization 2017). The current design is embedded in smartphones, which have very good privacy protection functions. The humour intervention itself is also very discreet. Given the popularity of smartphones among young people, this tool will help dispel their fears of formal intervention, so it is particularly important to target young people, since suicide is the second most common cause of death among 15–29-year-olds after road traffic accidents.

The multi-disciplinary nature of this humour design concept requires the involvement of researchers from different fields with different areas of expertise (Saariluoma et al. 2016), including cognitive science, neuroscience, psychology, sociology, physiology, health care, telecommunication and ICT. The research team should specialize in signal analysing, psychological experiment design, programming, health care service, cognitive intervention, app development and so on. The equipment used should be of the highest standard; it should be sophisticated, effective, easy to operate and very portable. Conducting mHealth research via mobile phones is at the cutting edge of health care research. Thus mHealth projects require realistic expectations and planning, integrating complementary sets of expertise in the research team, and an ability to remotely monitor, detect and flexibly resolve obstacles as they arise (Kane and Mohr 2016). This chapter explored ways to apply basic research findings to real life and to balance scientific study with daily technical artefacts. However, this process requires more specific in-depth researching, testing, re-researching, retesting and application.

5.8 Summary

This chapter employs a neuroscience approach to elaborate on the research results produced by joint studies within design science research, neuroscience and computer science. The guiding theories in this humour design concept are LBD and IKS, including investigations into the biological, psychological and social–cultural factors of potential users. It calls for holistic research on people before proposing a particular design concept. Humour plays an important role in business, politics, education and health care; it has the potential to relieve stress, promote a harmonious interpersonal environment and improve life satisfaction. Research on humour's positive influence on both physical health (such as the cardiovascular system, neuroendocrine system, immune system and pain tolerance) and mental health (as a coping mechanism and a strategy to enhance one's mood) drives the formation of a humour design concept built on LBD and IKS—the 6Rs. model.

The studies of EEG-based BCI provide more possibilities to apply this humour design concept to real lives. People who suffer from depressive mood are exemplified in this chapter to illustrate the method of conducting the investigations on LBD and IKS. To guarantee a successful design outcome, mHealth is integrated into the humour design concept as the medium. The design intention is to shift more attention and resources to preventive care and early intervention to help people detect problems earlier and overcome the damage caused by negative moods.

This chapter describes the humor design concept, highlighting the benefits of real-time recording, feedback and humor intervention. With the assistance of more sophisticated and advanced technical tools of BCI, biological signs, such as brain signals, heart rate, blood pressure, skin conductance and eye-tracking, can be recorded simultaneously or selectively in real time, and combined with subjective self-reports, researchers will achieve more objective and reliable evaluations of people's emotional states.

References

Adam MTP et al (2017) Design blueprint for stress-sensitive adaptive enterprise systems. Bus Inf Syst Eng. https://doi.org/10.1007/s12599-016-0451-3

Akerstedt T (1985) Adjustment of physiological circadian rhythms and the sleep-wake cycle to shiftwork. In: Hours of work temporal factors in work scheduling

Akiskol HS (2017) Mood disorders: clinical features. In: Kaplan and Sadock's comprehensive textbook of psychiatry

Bakker D et al (2016) Mental health smartphone apps: review and evidence-based recommendations for future developments. JMIR Mental Health e7. https://doi.org/10.2196/mental.4984 PM - 26932350 M4 - Citavi

Beck AT (1991) Cognitive therapy: A 30-year retrospective. Am Psychol. https://doi.org/10.1037/0003-066X.46.4.368

Beck J (2011) Cognitive behavior therapy: basics and beyond. The Guilford Press. https://doi.org/10.2307/1940491

Beck AT, Alford BA (2009) Depression: causes and treatment. J Chem Inf Model. https://doi.org/10.1017/CBO9781107415324.004

Ben-Zeev D et al (2013) Mobile technologies among people with serious mental illness: opportunities for future services. Administration Policy Ment Health Ment Health Serv Res. https://doi.org/10.1007/s10488-012-0424-x

Berkes F, Turner NJ (2006) Knowledge, learning and the evolution of conservation practice for social-ecological system resilience. Human Ecol. https://doi.org/10.1007/s10745-006-9008-2

Cann A, Calhoun LG (2001) Perceived personality associations with differences in sense of humor: Stereotypes of hypothetical others with high or low senses of humor. Humor—Int J Humor Res 14(2). https://doi.org/10.1515/humr.14.2.117

Caplan SE (2003) Preference for online social interaction: a theory of problematic internet use and psychosocial well-being. Commun Res. https://doi.org/10.1177/0093650203257842

Crawford SA, Caltabiano NJ (2011) Promoting emotional well-being through the use of humour. J Positive Psychol. https://doi.org/10.1080/17439760.2011.577087

Davidson RJ (1993) Cerebral asymmetry and emotion: conceptual and methodological conundrums. Cogn Emot. https://doi.org/10.1080/02699939308409180

Detera-Wadleigh SD et al (1998) Genome screen and candidate gene analysis in bipolar disorder. Am J Med Genet Neuropsychiatric Genet

Drevets WC (2001) Neuroimaging and neuropathological studies of depression: Implications for the cognitive-emotional features of mood disorders. Curr Opin Neurobiol. https://doi.org/10.1016/S0959-4388(00)00203-8

Falkenberg I (2010) Entwicklung von Lachen und Humor in den verschiedenen Lebensphasen. Zeitschrift für Gerontologie und Geriatrie. https://doi.org/10.1007/s00391-009-0085-x

Falkenberg I et al (2011) Implementation of a manual-based training of humor abilities in patients with depression: a pilot study. Psychiatry Res 186(2–3):454–457. https://doi.org/10.1016/j.psychres.2010.10.009 (Elsevier Ireland Ltd)

Forgas JP, Bower GH, Krantz SE (1984) The influence of mood on perceptions of social interactions. J Exp Soc Psychol. https://doi.org/10.1016/0022-1031(84)90040-4

Fredrickson BL, Joiner T (2002) Positive emotions trigger upward spirals toward emotional well-being. Psychol Sci 13(2):172–175. https://doi.org/10.1111/1467-9280.00431

Free C et al (2013) The effectiveness of mobile-health technology-based health behaviour change or disease management interventions for health care consumers: a systematic review. PLoS Med 10(1) https://doi.org/10.1371/journal.pmed.1001362

Freud S (1960) Jokes and their relation to the unconscious, vol 6. In: VIII of the standard edition of the complete psychological works of Sigmund Freud

Gervais M, Wilson DS (2005) The evolution and functions of laughter and humour: a synthetic approach. Q Rev Biol. https://doi.org/10.1086/498281

Giese MA, Poggio T (2003) Cognitive neuroscience: neural mechanisms for the recognition of biological movements. Nat Rev Neurosci. https://doi.org/10.1038/nrn1057

Harmon-Jones C, Beer JS, Harmon-Jones E (2009) Collaborations in social and personality neuroscience. Methods Soc Neurosci

Henriques JB, Davidson RJ (1991) Left frontal hypoactivation in depression. J Abnorm Psychol. https://doi.org/10.1037/0021-843X.100.4.535

Heron KE, Smyth JM (2010) Ecological momentary interventions: Incorporating mobile technology into psychosocial and health behaviour treatments. Br J Health Psychol. https://doi.org/10.1348/135910709X466063

Hird N, Ghosh S, Kitano H (2016) Digital health revolution: Perfect storm or perfect opportunity for pharmaceutical R&D? Drug Discov Today 21(6):900–911. https://doi.org/10.1016/j.drudis.2016.01.010 (Elsevier Ltd)

Kane JM, Mohr DC (2016) Intervention studies 42(2):157–167. https://doi.org/10.1007/s10488-014-0556-2.Strategies

Kaslow NJ et al (2007) Health care for the whole person: research update. Professional Psychol Res Pract. https://doi.org/10.1037/0735-7028.38.3.278

Kazdin AE, Blase SL (2011) Rebooting psychotherapy research and practice to reduce the burden of mental illness. Perspect Psychol Sci. https://doi.org/10.1177/1745691610393527

Kessler RC et al (2001) The use of complementary and alternative therapies to treat anxiety and depression in the United States. Am J Psychiatry. https://doi.org/10.1176/appi.ajp.158.2.289.

Laceulle OM et al (2015) Adolescent personality: associations with basal, awakening, and stress-induced cortisol responses. J Pers. https://doi.org/10.1111/jopy.12101

Leahy RL, Dowd ET (2002) Clinical advances in cognitive functioning. https://doi.org/10.1177/0956797612450893

Leikas J (2009) Life-based design. A holistic approach to designing humantechnology interaction

Lewinsohn PM et al (1993) Adolescent Psychopathology: I Prevalence and incidence of depression and other DSM-III-R disorders in high school students. J Abnormal Psychol. https://doi.org/10.1037/0021-843X.102.1.133

Li M, Xu H, Lu S (2018) Neural basis of depression related to a dominant right hemisphere: a resting-state fMRI study. Behav Neurol. https://doi.org/10.1155/2018/5024520

Martin R (2010) The psychology of humor: an integrative approach

McCarley JS, Kramer AF (2009) Eye movements as a window on perception and cognition. In: Neuroergonomics: the brain at work. https://doi.org/10.1093/acprof:oso/9780195177619.003.0007

Meeker M, Wu L (2013) 2013 internet trends, Kleiner Perkins Caufield byers insights

Moran CC (1996) Short-term mood change, perceived funniness, and the effect of humor stimuli. Behav Med. https://doi.org/10.1080/08964289.1996.9933763

Mykles DL et al (2010) Grand challenges in comparative physiology: integration across disciplines and across levels of biological organization. Integr Comp Biol. https://doi.org/10.1093/icb/icq015

Nestler EJ et al (2002) Neurobiology of depression. Neuron 34(1):13–25. https://doi.org/10.1016/S0896-6273(02)00653-0

Nezlek JB, Hampton CP, Shean GD (2000) Clinical depression and day-to-day social interaction in a community sample. J Abnorm Psychol. https://doi.org/10.1037/0021-843X.109.1.11

Oulasvirta A et al (2012) Habits make smartphone use more pervasive. Pers Ubiquit Comput. https://doi.org/10.1007/s00779-011-0412-2

Parasuraman R (2008) Neuroergonomics analyzing brain function to enhance human performance in complex systems Raja Parasuraman. Security (December 2008)

Parasuraman R, Rizzo M (2009) Neuroergonomics: the brain at work. https://doi.org/10.1093/acprof:oso/9780195177619.001.0001

Pelphrey KA et al (2005) Functional anatomy of biological motion perception in posterior temporal cortex: An fMRI study of eye, mouth and hand movements. Cereb Cortex. https://doi.org/10.1093/cercor/bhi064

Proctor CL (2017) Positive psychology interventions in practice. Positive Psychol Interv Pract. https://doi.org/10.1007/978-3-319-51787-2

Proudfoot J (2013) The future is in our hands: the role of mobile phones in the prevention and management of mental disorders. Aust New Zealand J Psychiatry 47(2):111–113. https://doi.org/10.1007/s10488-012-0424-x, https://doi.org/10.1177/0004867412471441

Rickard N et al (2016) Development of a mobile phone app to support self-monitoring of emotional well-being: a mental health digital innovation. JMIR Mental Health. https://doi.org/10.2196/mental.6202

Rissanen T (no date) Studies on life satisfaction in samples of the general population and depressive patients

Rousi R et al (2011) Life-based design as an inclusive tool for managing microinnovations. In Lecture notes in informatics (LNI), Proceedings—series of the Gesellschaft fur Informatik (GI)

Ruch W, McGhee PE (2014) Humor intervention programs. In: The Wiley Blackwell handbook of positive psychological interventions. https://doi.org/10.1002/9781118315927.ch10

Saariluoma P, Leikas J (2010) Life-based design—an approach to design for life. Global J Manag Bus Res

Saariluoma P, Cañas JJ, Leikas J (2016) Life-based design. Available at: https://link.springer.com/chapter/10.1057/978-1-137-53047-9_6

Shen N et al (2015) Finding a depression app: a review and content analysis of the depression app marketplace. JMIR mHealth and uHealth 3(1):e16. https://doi.org/10.2196/mhealth.3713

Steger MF, Kashdan TB (2009) Depression and everyday social activity, belonging, and well-being. J Counsel Psychol. https://doi.org/10.1037/a0015416

van Bilsen H (2013) Cognitive behaviour therapy in the real world: back to basics

Wang Y, Wang H, Ye X, Lei Z (2020) Simulation of individual knowledge system and its application. J Syst Sci Syst Eng. https://doi.org/10.1007/s11518-019-5452-6

Wheeler RE, Davidson RJ, Tomarken AJ (1993) Frontal brain asymmetry and emotional reactivity: a biological substrate of affective style. Psychophysiology. https://doi.org/10.1111/j.1469-8986.1993.tb03207.x

World Health Organization (2017) Global Hepatitis report, 2017. World Health Organization. https://doi.org/10.1149/2.030203jes

World Health Organization and WHO, W. H. O. (2012) World malaria report, world malaria report 2011. ISBN 978 92 4 1564403

Zhu MY et al (1999) Elevated levels of tyrosine hydroxylase in the locus coeruleus in major depression. Biol Psychiat. https://doi.org/10.1016/S0006-3223(99)00135-3

Part III
Emotions, Culture and Aesthetics

Human values are formed through complex cultural, social, cognitive, and emotional processes. Culture, can be seen as playing a major role in shaping human values and their interpretation, and subsequently framing the ways in which phenomena are emotionally experienced according to these values and understandings. Aesthetic appreciation, and the experience of what is pleasant or beautiful, is tightly intertwined with culture, its definitions of beauty and pleasantness, its positioning of relationships, as well as its explication and enactment of value. Modern business for instance, thrives of culturally constructed value through brand building and corporate image and establishing connections between products and services with social values and lifestyles. Aesthetics exist within the feeling or emotional connection humans have with technology design. They have material qualities that manifest in symbolism, code and physical sensations that are then de-coded within the feeling or sense-making processes of experience, to generate the subsequent emotional reaction(s) that remain in re-call, re-use and re-experience. Through tapping into culture via cross-cultural, historical and high cultural perspectives, we are able to unravel the ways in which people are programmed (cultivated, or culturally conditioned) to emotionally experience designed phenomena in the ways they do. The next few chapters deal precisely with these issues. They observe various domains of design and cultural production and examine, observe and philosophized on the connections between culture, aesthetics and emotional experience.

Chapter 6
The Good, the Bad and the Ugly Graffiti

Mari Myllylä

Abstract Emotions play an essential role in aesthetic and art experience. Graffiti is an example of urban visual communication, and it can also be understood as a form of art. Like other works of art, graffiti can evoke different aesthetic emotions in its audiences, such as pleasure, wonder, interest and pride but also disinterest, disappointment or embarrassment, and even anger and disgust—further impacting, for example, how they value this art form. However, few studies have explored what kinds of emotions people feel when they appraise graffiti. This chapter discusses emotions in graffiti using examples from participant interviews in the Purkutaide study. Interview quotes are assessed against theories regarding aesthetic emotions and art appreciation. There are several challenges associated with studying emotions inspired by graffiti. For instance, explicating emotions verbally is difficult, and the same graffiti work can be interpreted as beautiful or ugly, or good or bad, depending on multiple factors. Appraising graffiti is an interactive and iterative process that depends on both the perceived visual and non-perceivable symbolic features of the work. The sociocultural and physical context, viewing time, subjective motives, the work's relation to the self, the level of learned graffiti-related expertise and other aspects may also influence what kinds of emotions graffiti evokes, and how it is judged in terms of good/bad or beautiful/ugly.

6.1 Introduction

Each individual has his or her own unique mental representations of 'graffiti'. These representations are based on, for example, previous personal histories and life experiences, knowledge, social circumstances, incentives and even physical bodily interactions. Our experiences are often, if not always, coloured by an array of felt emotions that both affect (and are affected by) how we perceive, evaluate and value graffiti.

M. Myllylä (✉)
Faculty of Information Technology, University of Jyväskylä, P.O. Box 35 (Agora), FI-40014, Jyväskylä, Finland
e-mail: mari.t.myllyla@student.jyu.fi

© Springer Nature Switzerland AG 2020
R. Rousi et al. (eds.), *Emotions in Technology Design: From Experience to Ethics*, Human–Computer Interaction Series,
https://doi.org/10.1007/978-3-030-53483-7_6

87

Fig. 6.1 Work 1. Photo by Jouni Väänänen 2016

Possible incentives and typical characteristics of graffiti writers have been the focus of much graffiti-related research. Various meanings of graffiti among the graffiti writers[1] themselves, and the consequences of graffiti for the individual and the surrounding social environment, have been a source of discussion and debate since the emergence of contemporary graffiti in the late 1960s and early 1970s (Avramidis and Tsilimpoudini 2017). However, the emotions that graffiti can elicit, especially among the people who view and experience it, have been largely overlooked.

This chapter discusses the role of emotions in appraising graffiti, based on the preliminary findings of a study that investigated the perceptions, evaluations, thoughts and emotions it evokes. This study was conducted in 2016 during a Purkutaide project (Purkutaide 2019), where an empty building scheduled for demolition, previously used as a business premises, was painted inside and outside with legal and commissioned graffiti and murals. Purkutaide project aims to use empty real estates that are in the end of their lifecycles for art and other related activities. This non-profit project started in Kerava in 2016, where 106 different artists created graffiti and mural works covering about 4850 m[2] of interior and external surfaces (Purkutaide 2019). I interviewed 19 participants, from laypeople to graffiti writers. I used semi-structured interviews in a thinking-out-loud method to record participants' self-reports while they studied four selected works of graffiti (Figs. 6.1, 6.2, 6.3 and 6.4) and one mural painting inside the building (this article focuses only on the answers related to graffiti works 1–4).

Each work was perceived and assessed by one person at a time, stopping at one work and then the next. One of the questions specifically asked about emotions is:

---

[1] I am deliberately using the term 'graffiti writer', because in the graffiti vernacular the graffiti production is typically referred to as 'writing' and its producers as 'writers', instead of painters or artists.

Fig. 6.2 Work 2. Photo by Jouni Väänänen 2016

Fig. 6.3 Work 3. Photo by Jouni Väänänen 2016

'How do you feel when you look at this work? What kind of feelings does it evoke in you?' Participant comments from that study are used in this article to illustrate examples of possible emotions elicited by selected graffiti (art) works, and how they relate to existing models and suggestions about aesthetic emotions. The term 'aesthetics' can have several meanings, from its broader connotation of 'philosophy of art' to the narrower 'sense perception' or 'sensory cognition' of a subject who is interacting with an artwork (Carroll 1999). It can also be used as an adjective conjoint

Fig. 6.4 Work 4. Photo by Jouni Väänänen 2016

to a noun, such as 'aesthetic experience' or 'aesthetic attitude', referring to a special contemplative mental state that occurs in response to an object (Carroll 1999). In this article I mainly use the third definition. I translated the participant comments from the original Finnish language to English.

Because of the complexity, difficulty and lack of sufficient research, not all aspects of emotions in graffiti can be reviewed here. For example, I do not discuss the emotions involved in creating graffiti. How graffiti are assessed and appreciated is similar to how artworks are evaluated—that is, not only according to their apperceived aesthetic or artistic worth, but based on a multitude of moral and other values, grounded in emotions and emotional responses (Fingerhut and Prinz 2018). Thus, the concept and appraisal of graffiti can (and does) evoke an array of different and even opposite emotions, not only impacting whether it is valued as 'good' or 'bad', but also colouring and further amplifying some deeply personal opinions, judgments, and rational or irrational-seeming behaviour.

6.2　Definitions

The definition of graffiti varies widely depending on the researcher and discourse (see, for example, Ross 2016a). I use the definition provided by Ross (2016b, 476): '[Graffiti] typically refers to words, figures, and images that have been written, drawn and/or painted on, and/or etched into or on surfaces.' It ranges from tags (simple and quickly written pseudonyms of the graffiti writer) to throw-ups (large sprayed bubble letters) to pieces (expert work with colourful, detailed and complex letters and images).

Tools to produce graffiti can vary from marker pens to spray paint and even fire extinguishers. Graffiti writers also have special aesthetic hand styles (Ross 2016b), which distinguishes their artistic style from other visual outputs and aesthetic genres, such as murals or other forms of urban art. Graffiti is typically done without permission, but in its modern form, sometimes referred to as 'post-graffiti', it can also be done legally, transforming it from 'illegal urban action to a legal canvas art' (Ross 2016b, 477).

6.2.1   Graffiti as Communication and Art

In its elementary essence, graffiti can be considered as a form of visual communication (Brighenti 2010; Wacławek 2011; Young 2005). It is also a cultural artefact: products of the graffiti subculture have their own rules, norms, hierarchy system and even language (Campos 2012). Graffiti has been described as urban folk art (Ferrell 2017), urban art (Austin 2010; Valjakka 2016) and a post-modern art form (Dempsey 2003). Art is also a form of communication (Dewey 2005), and artworks can be seen as 'communicative devices' (Seeley 2015, 23), conveying emotional information via signalling codes, classifiers and modifiers to determine the logical–semantic hierarchy of the message, which provides the viewer with reasoning alternatives (Gombrich 1963).

Graffiti is not art a priori, though; it can either be art or not. According to Solso (2003, 15): 'Art is a perception consciously experienced and defined by human beings as aesthetic.' For something to be considered art, it also needs to be interpreted in as being representational and/or symbolic (Solso 2003). However, whether a specific graffiti work can be defined as 'art' is not only based on the work's visual features and the perceiver's personal taste—aspects that humans commonly experience as aesthetic; it also depends on how it is agreed and fostered in its historical, sociocultural discourse, between individuals, groups and institutes (Kimvall 2014; Myllylä 2018). Similar factors that influence whether graffiti is considered or felt as art or not, or as beautiful or ugly or something else, provides an interesting context for investigating emotions in graffiti.

6.2.2   Emotions

In order to be able to describe what kinds of emotions graffiti can elicit, it is necessary to first clarify what is meant by emotions. Emotions may be understood as temporary mental episodes that are internal states, or unconscious and automatic recursive processes, which are adaptive responses to external events and features and their appraised importance for the organism (Frijda 2008; Moors et al. 2013; Silvia 2005a; Solso 2003). According to a componential view of emotion, emotional episodes consist of five subsystemic components that evolve and provide feedback to each other in conjoint coordination during an emotional episode (Meuleman et al. 2019; Moors et al. 2013). These components include: (1) cognitive appraisal, for evaluating a stimulus and interacting with the environment in reflection of their subjective significance; (2) a motivational component, related to behavioural action tendencies and readiness; (3) a somatic or physiological component related to changes in brains and autonomic and peripheral bodily responses; (4) a motor or expression component for changes in involved behaviour and, for instance, facial and vocal expressions; and (5) a subjective feeling component for integrating all the former into a 'gestalt' experience, which may be categorised as or generate a verbal output such as a certain labelled feeling (Meuleman et al. 2019; Moors et al. 2013; Silvia 2009). Baumeister et al. (2007) describes emotion as 'a state of conscious feeling, typically characterized by physiological changes such as arousal' (Baumeister et al. 2007, 168–169). An emotion may be experienced as a single state, but it is often blended with several other emotions and moods, and runs in parallel with several other emotions or emotional episodes (Moors et al. 2013).

The concept of basic emotions frequently emerges in discussions of emotions. According to Izard (2007, 261), basic emotions have 'evolutionarily old neurobiological substrates, […] an evolved feeling component and capacity for expressive and other behavioral actions of evolutionary origin'. Such emotions are prompted quickly, automatically and unconsciously when a person senses or perceives a stimulus that activates evolutionary-based neural and mental processes, leading to stereotypical responses that are each associated with unique feelings. Basic emotions do not require higher-level, complex cognitive appraisals, such as thinking or judgment. However, these emotional responses can change and be regulated as a result of both learning new knowledge and because of the development of a person's information processing and motor activity capabilities (Izard 2007).

There is no agreement on what exactly these basic emotions are. Izard (2007) defines them as 'interest, joy/happiness, sadness, anger, disgust, and fear' (Izard 2007, 261). According to Ekman (1992, 1999), basic emotions include anger, awe, contempt, disgust, embarrassment, excitement, fear, guilt, interest, sadness, shame, surprise, enjoyment (from sensory sources and of accomplishment), amusement, contentment, relief, pride in achievement and satisfaction. Panksepp (2006) sees basic emotions as lust, care, panic, play, fear, rage and seeking.

In contrast to conscious emotions, which 'stimulate reflection and learning' (Baumeister et al. 2007, 170), affect can be defined as an automatic (either conscious or unconscious) response to a stimulus—a quickly arising and simple feeling of something to be approached or avoided, liked or disliked. It is a less intense feeling than emotion, and it might not be linked to a physiological arousal. Parallel affects can arise out of perceiving something and associating it as good or bad; they are thus simple reactions (Baumeister et al. 2007). Clore and Ortony (2008, 629) view emotions as 'cognitively elaborated affective states' that include multiple representations of something being good or bad at the same time. Whether conscious emotions and affects are the same or separate phenomena, they have a deep impact on a person's further cognising, bodily functions and behaviour, as they direct interactions involving, for example, perception and attention, judgements, values, learning, memory, goals, motivational priorities, categorisation and conceptual frameworks, physiological reactions, communication processes, estimates and situational assessments (Tooby and Cosmides 2008).

Our minds and bodies work together and affect each other: an emotional experience can be moulded by the individual's biological state, such as fatigue or hunger, as well as unique features of an individual's perceptual systems such as vision or hearing, attention and its limitations, gender, age and perhaps even (emotional) intelligence, and many other reasons. Evaluations can also be affected by the real or imagined presence of others and physical and mental interactions, which include the viewer's own body and its movements, perceived objects and events, and other people. As Colombetti (2010) notes, assessments and appraisals arise in a situated organism in a specific bodily state of arousal. Making sense of events can be seen as embodied, cognitive–emotional understanding (Colombetti 2010).

6.2.3 Emotions and Appraisal

Clore and Ortony (2008) suggest that emotions are implicitly about something being good or bad, and they need to be evaluated somehow. Appraisal theories explain the emergence of emotions as a process, in which initial affective reactions are constructed into emotions in an iterative and recursive appraisal process (Clore and Ortony 2008; Cunningham and Zelazo 2007; Moors et al. 2013); different emotional states are 'refined, situated, further evaluated, and rerepresented' (Clore and Ortony 2008, 639), resulting in versatile emotional states or emotional episodes. According to Cunningham and Zelazo (2007), in each iteration, information from the previous 'cycle' is conveyed between higher- and lower-order processes, recalculated and shaped further as new information and attitude representations are included in the evaluation. The number of iterations may depend on variables such as individual abilities, motivation, and the available resources and opportunities to conduct the appraisal process (Cunningham and Zelazo 2007).

Emotions and cognition are intertwined, and emotions often emerge as a result of cognitive evaluation or appraisal, reflecting how an event or outcome relates to a person's subjective needs, values, motives, beliefs, current goals and other concerns, assuming it seems to make sense, matters, and is relevant to that person and their wellbeing (Baumeister et al. 2007; Moors et al. 2013; Silvia 2005a, b; Thompson and Stapleton 2009). According to Clore and Ortony (2008), the emotional appraisal process has two parts: an associative aspect that is based on prior subjective experience, similar to other situations and temporal contiguity, and a slower rule-based reasoning based on the individual's developed ability to make computational distinctions. These two properties of appraisal ensure that the individual is prepared to react to fast events and has the flexibility to ensure the correctness of their emotional estimates (Clore and Ortony 2008).

The cognitive appraisal process creates inputs for new emotional outcomes, which can depend, for example, on the time available for processing and the amount of the recursive appraisal cycles—for instance, related to the viewing time of an artwork, or what kind of emotional output a person has learned to anticipate from a certain behaviour (Baumeister et al. 2007; Brieber et al. 2014; Cunningham and Zelazo 2007; Moors et al. 2013; Tinio and Gartus 2018). A person's pre-existing attitudes and values, together with their current goals and information about the stimulus and context—such as background information about the artwork and the artist—may also affect how they appraise the object's valence, whether something is good or bad, and how further actions are planned (Cunningham and Zelazo 2007; Fingerhut and Prinz 2018; Gerger et al. 2014; Tinio and Gartus 2018). As Leder and Nadal (2014) suggest, appraisals are the key mechanisms to elicit aesthetic and art-related emotions and experiences.

6.2.4   Emotions and Communication

Expressing emotions and understanding the emotions of others is crucial for humans as a social species (Solso 2003). Emotions are not only internal or subjective experiences; they also function as communication when an individual interacts with her social and physical environment (Baumeister et al. 2007; App et al. 2011). According to App et al. (2011), emotions can be expressed and understood in different non-verbal channels, and specific channels seem to be optimised for specific types of emotions. Different emotional displays, such as facial expressions, body movements or certain types of touch, seem to be fine-tuned to communicate certain emotional messages, thoughts and intentions. They are therefore important for coordinating different aspects of an individual's life, such as social status and intimate relationships (App et al. 2011).

Also, as Baumeister et al. (2007) notes, the emotions of one person may influence the actions and emotions of other people, and people may behave in certain ways in anticipation that this will elicit certain feelings and emotions in others. Outward expressions may therefore not always correspond to a person's subjective emotional experience (App et al. 2011). According to Tooby and Cosmides (2008), a person can regulate her emotional expressions and share only that emotional information with others she sees as beneficial, depending, for example, on what kind of relationship the person has with the receiver of that information, or whether she is alone or with people who have similar or opposite interests. Some of the underlying mechanisms for regulating emotional expression can be innate and unconscious, and some may depend on individual development as well as cultural and social learning (Ekman 1999; Tooby and Cosmides 2008).

Since people can modulate their emotional expressions, and not all emotions are easily expressed verbally, there are challenges related to using self-reported data on emotions and emotional experiences. As Barrett (2006) notes, even though verbal self-reports about emotions can be more about the use of language related to emotions than the emotional experiences themselves, they at least give some information about the emotional experience, the valence of affective categories such as feeling pleasant or unpleasant, and high or low arousal states of the individual. However, such self-reported information seems to reveal more about affective states than distinct emotional categories (Barrett 2006). As Frijda (2008, 37) argues, emotions can generate many different feelings in diverse ways and modes, 'reportable or not reportable, diffuse and global or articulate and amenable to verbal description', which makes it difficult to research conscious emotions.

6.2.5 Aesthetic Experience, Art Appreciation and Emotions

An aesthetic experience may be understood as the result of a complex and ongoing interplay among multiple perceptual, cognitive and emotional processes that can cause a variety of simultaneous and even contradictory emotions (Gartus and Leder 2014; Leder and Nadal 2014). An aesthetic experience has sometimes been called a distinctively aesthetic state of mind that is different from, for example, a religious or cognitive state, and which serves as a basis for explaining 'aesthetic properties, qualities, aspects, or concepts' of the aesthetic object, judgement and value (Iseminger 2005, 2). An aesthetic state of mind is different from a sensual pleasure; it does not require prior ideas about or concepts of art, and it can be focused on both art and non-art (Iseminger 2005).

Art and aesthetic experience are closely related concepts, but they do not necessarily go hand in hand. Although some art philosophers assert that the function of art can be described as a vehicle to afford aesthetic experiences, art may be understood as a stricter domain focusing on art objects instead of a broader concept of aesthetics, which is more about a response to any sources of aesthetic experience (Carroll 1999). Many mundane and everyday actions, such as cleaning the house, can give rise to an aesthetic experience (Dewey 2005). In the case of graffiti, an individual may experience it as aesthetically pleasing or not, and at the same time evaluate it as 'artistic' or not, regardless of whether they would consider the graffiti a 'work of art'.

According to Dutton (2009), art appreciation arises from the imagination and direct pleasure generated by the perceived object, which is related to the work's recognisable styles and the demonstration of technical skills, virtuosity and the artist's creativity. Artistic creations are expressions of individual personality: they are saturated with emotions, challenge their creators and perceivers intellectually, and induce pleasure when those challenges are solved (Dutton 2009). Fingerhut and Prinz (2018) propose that when those aesthetically praised features are present, their artistic goodness, which leads to art appreciation, is seized by the emotion of wonder. Wonder can be generally characterised as a positive emotion that may be cognitively baffling and ambivalent, perceptually captivating, and create a sense of appreciation and respect, engaging us to further appraise artwork, invest our resources into exploring wonderous experiences, and enable thinking styles that promote tolerance for uncertainty and openness to new possibilities (Fingerhut and Prinz 2018).

The aesthetic experience and artistic evaluations of a work may be influenced by the viewer's individual characteristics such as attitudes, interests and knowledge regarding, for example, art styles and art movements (Gartus and Leder 2014). However, as Gartus and Leder (2014, 447) note, we may be 'emotionally moved by artworks we understand poorly, and it is possible to feel indifferent towards artworks we understand well and judge highly'. A whole set of priming factors affects every aesthetic experience: the social discourse and its prejudices, expectations and aesthetic orientations, as well as the context and situation, all shape the anticipations and define the environmental prerequisites for assessing an object (Gartus and Leder 2014; Gerger et al. 2014; Leder and Nadal 2014). For example, if an individual thinks she is perceiving a work of art instead of a photograph of real events, this may change how she relates to the work, as well as her judgements and emotional reactions (Van Dongen et al. 2016). Even an individual's personality can affect their aesthetic experience and judgements; for example, openness to new experiences can have a positive effect on art and aesthetic appreciation (Fayn and Silvia 2015; Gartus and Leder 2014).

Although many disparities in art have evoked emotions between individuals, it may be possible to find some clusters of emotion types that each specific artwork typically evokes in most of its viewers (Tinio and Gartus 2018). As Tinio and Gartus (2018, 338) suggest, even though there are individual-level differences between people, we all share the same biological similarities and respond to certain artworks' 'aesthetic emotional affordances' in a similar, common fashion. Seeley (2015) describes aesthetic emotion as the result of a reduction in ambiguity of an evaluated artwork via cognitive mastering, where success in classification and evaluation generates an emotional state of pleasure or satisfaction (Seeley 2015). Simple feelings of liking or disliking, preference and pleasure from art are important, because 'much of human experience is simple and mild' (Silvia 2009, 48). However, art can also evoke more complex, special emotions such as beauty, pleasantness, interest and surprise, awe and chills, and even negative emotions such as anger, disgust, shame and embarrassment (Fayn and Silvia 2015; Silvia 2009). These kinds of emotions are often mentioned in graffiti and street art-related discussions (Dickens 2008; Halsey and Young 2006; Taylor 2012; Young 2005).

6.3   Aesthetic Emotions in Graffiti

Research on aesthetic evaluations has often focused on the central themes of positive/negative dimensions of beautiful/ugly or appealing/not appealing (Fayn and Silvia 2015). Another way to approach special aesthetic emotions is to group them into higher-level categories such as knowledge, hostile and self-conscious emotions (Silvia 2008).

6.3.1 Knowledge Emotions

'Knowledge emotions' include interest, confusion, surprise and awe. They are related to goals and associated with learning (Silvia 2010). Such emotions are appraised based on an event's novelty and complexity, which can include assessing something as new, surprising, unexpected or mysterious; and its comprehensibility, a sort of a coping potential in which a person assesses whether she has the necessary knowledge and skills to cope with and understand an event or object (Silvia 2008).

Awe, which can be understood as 'a term for intense wonder' (Fingerhut and Prinz 2018), refers to something experienced as extraordinary, special, vast, physically or mentally larger than oneself or mundane everyday life (Fayn and Silvia 2015; Fingerhut and Prinz 2018). The importance of awe and wonder emotions is implied in the Purkutaide study:

> I have seen so much graffiti that it must be at some level really exceptional for it to evoke any passion. Any graffiti piece is good merely because it exists, but it has to have something that lifts it above others, that it erodes into deeper consciousness. (Graffiti writer, over 40 years old)

In order to evoke strong emotions and awe, the artwork needs to be exceptional or somehow special compared to others. When an event or object is new and complex, it is typically considered interesting, but once it loses this novelty, interest may be lost (Silvia 2010). Like artwork, graffiti may also contain hidden and unknown elements that are appraised so that they evoke emotions of mystery and even excitement:

> This piece reflects something similar mysticality and the character is hidden by a mask, it evokes a criminal feeling, what graffiti basically has been. Something a little bit of criminal and exciting. (Knows some about graffiti, 20–30 years old)

However, there is a fine line between experiencing something as positively intriguing and being negatively affected by not knowing anything about it. Being mysterious may evoke positive excitement and interest, but a lack of knowledge may also generate uncertainty and even fear:

> This is a little bit scary. I see that here a story continues in a western style from left to right, I can see the characters' direction going that way, but where does this go? I should know more about this. (Knows some about graffiti, over 40 years old)

In addition to being an aesthetic emotion, interest is also a basic emotion that occurs throughout a person's life, responding to 'novelty, change, and the opportunity to acquire new knowledge and skills' (Izard 2007, 264). If the work does not have such properties to interest the appraiser, it may cause flat emotions and even disappointment:

> I feel a bit of a disappointment, not really anything else. It is neutral and like a wallpaper. It does not offend anyone, it just is. (Graffiti writer, over 40 years old)

When there is no interest, there may also be a lack of strong emotions; the artwork may just 'exist' in a neutral emotional space. Also, different things interest different people for different reasons:

> First I get a feeling that I think it is nice that the gang does these kinds, it is really pleasant that there are guys who do with spray paint something totally different from normal, but then again at the same time it is not my thing. I do not experience this work as very interesting, so this does not evoke any strong feelings in me in general. I pass these kinds quite quickly. (Graffiti writer, 30–40 years old)

If the work is not perceived as 'being my thing' or as something that would relate to the perceiver's own goals, it may be judged as disinteresting. In some cases, interest may rise because of personal memories or goals, or the work may have other personally meaningful content (Tinio and Gartus 2018), or be closely related to self-conscious emotions.

Comprehension and more knowledge may make the artwork appear more interesting (Silvia 2008, 2010; Tinio and Gartus 2018), and positively affect the emotional valence. As a person gains new knowledge and understands more complex concepts, she starts to 'see subtle differences and contrasting perspectives that aren't apparent to novices' (Silvia 2008, 59), which also affects emotional appraisals regarding art (Fayn and Silvia 2015; Kuuva 2007; Leder et al. 2004; Pihko et al. 2011; Silvia 2008). In the case of graffiti art, expertise in graffiti can also impact appraised emotions (Gartus and Leder 2014; Gartus et al. 2015). Some experts have found that it is possible to express suppressed emotional reactions and to approach artwork in a more emotionally detached style, where the focus and content of the experience is on the artwork's stylistic, formal and contextual properties (Leder et al. 2014). Similar suggestions can be found in the Purkutaide study, where graffiti writers— that is experts—generally seemed to focus and explain things related to the visual appearance of the work and how it would 'fit' into their standards, personal taste and own graffiti writing. However, verbally explicating emotions seemed to be difficult for everyone, from novices to experts.

The viewing time may affect the graffiti appraisal process, as it may be understood as iterative cycles in which each cycle produces new combinations of thoughts and emotions (Brieber et al. 2014; Moors et al. 2013; Tinio and Gartus 2018). In this way, viewing time may impact the comprehension of the appraised graffiti:

> It is a bit ugly, yes. Maybe now when I start to look at it, when I have just gazed at it when passing by and as part of a whole, when now staring at this more it begins to look finer, one focuses on that. Before I interpreted this as uglier than now. (Knows some about graffiti, 20–30 years old)

Viewing time may generate new and even opposite emotions, and impact how graffiti is judged and valued. It may require that the individual is voluntarily and deliberately putting effort and resources into the appraisal process.

6.3.2 Hostile Emotions

Some people may perceive graffiti as ugly, less skilled, unaesthetic, visual litter or vandalism, evoking negative feelings such as disgust or repulsion, uncontrolled and harmful activity caused by social outcasts, neglecting or discarding their aesthetic and artistic values (Young 2005). According to Silvia (2009), 'hostile emotions' include anger, disgust and contempt, and are experienced when an event is appraised as contrary to a person's own goals and values, as deliberately eliciting anger, or when something is appraised as unpleasant, harmful or dirty, and thus elicits disgust. Hostile emotions motivate aggression, violence and self-assertion (Silvia 2009). Some hostile emotions and assessing something as ugly were found also in the Purkutaide study:

> This is the ugliest or one of the ugliest of all these works. First is that character of course. It is probably some character, that is known in the graffiti circles, but for me it is just a blob, I don't know what it is. Then is the text, it does not really pop out to my eye. The whole thing is so garish, that even colour wise it does not pop out. I cannot make sense what it reads […] a bit unpleasant looking, where ooze is dripping. (Knows some about graffiti, 20–30 years old)

Interestingly, the same work may have evoked hostile emotions in some participants but appraised as beautiful, good, or even playful and joyous in others. This may depend on how an individual recognises and associates perceived content in her subjective contextual level; a character may be associated with either revolting slimy nonsense or a funny figure from one's childhood, generating disgust in the former and happiness in the latter. However, in most Purkutaide study cases where the works did not please the participants, they expressed their emotions as disinterest, lame or neutral, instead of having any strong negative emotions.

6.3.3 Self-Conscious Emotions

Silvia (2009) describes 'self-conscious emotions' as complex and consisting of pride, shame, guilt, regret and embarrassment. Such emotions are experienced when events are appraised as congruent or not with a person's own goals, values and self-image, when things are assessed as caused by a person herself or when events seem to be consistent or inconsistent with a person's own or cultural standards. Self-conscious emotions can be also collective and experienced in response to other people's behaviour, actions and achievements (Silvia 2009). 'A creator can be proud of a great piece of work, and the creator's family, friends, and fans can be proud, too', as Silvia (2009, 50) notes. Also, there may be something that the person can subjectively relate to in the perceived artwork or graffiti (Tinio and Gartus 2018). In the Purkutaide study, several participants expressed these kinds of self-conscious emotions, in both positive and negative terms. For example, a work may be appraised as pleasing due to its aesthetics but also because it is somehow assessed as similar to the appraiser's own artistic practice:

> Even though I have said many times that the aesthetic part is secondary, I am now saying that this pleases me personally the most because of its style and composition and everything. Maybe exactly because this style of work I have done myself too lately, that there is some subconscious connection to my own doing. (Graffiti writer, over 40 years old)

It is easy to admit that noticing similarities to one's own goals, standards and physical activities may evoke positive emotions such as pride and feelings of a mental connection to the artist. However, if the work does not meet the expectations and collective standards of the participant, it may cause mixed emotions, where disappointment can be read between the lines:

This is confusing, so bafflement is probably the emotion. I know that [the graffiti creator name] is a skilled painter and can do a lot of things, so I would say it leaves me a bit empty [...] For me it is difficult to see anything more in this. In a way it is cheerful and perky [...] but as a work it does not leave me with a joyful feeling. (Knows some about graffiti, 30–40 years old)

As the previous extract suggests, being aware that one is appraising artwork that is expected to meet certain subjective criteria, and that should generate at least some positive emotions, can create an emotional collision with the pre-expectations and the experienced results of the appraisal if these expectations are not met. Inconsistencies between expectations and the actual experience may leave a person with disappointment and 'empty' or flat emotions, even though in theory (at a subjectively aware cognitive level) some visually perceivable elements of the work would suggest otherwise.

6.3.4 Emotions Related to Being Ugly or Beautiful

In the Purkutaide study, many participants found it difficult to verbalise their emotions. Instead, they generally first identified feelings with a positive or negative valence: the graffiti was either liked or disliked. In some cases, there was an emotion related to the work (such as interest), without the responder being able to define the work as beautiful or not. In most cases, instead of categorising graffiti as beautiful or ugly, it was instead evaluated as neither or both, or as 'nice', 'stylish', 'fine', 'quite beautiful' or 'pleasant'. When effort was put into exploring the details of a piece of graffiti, an individual may be positively moved and be able to describe certain perceivable features of the work, such as its technical and stylistic execution:

This is perhaps quite calming, even though there are a lot of cutting forms, still this is constructed as a balanced whole. This is enormous [...] but the colour scheme is very balanced or very simple [...] However this is not by any means boring, the dimensions and forms and cuttings of the letters come well to the fore. (Knows some about graffiti, 30–40 years old)

Shapes, colours, forms and other perceivable properties can make the work appear visually balanced and interesting. However, observing certain balanced visual qualities in a work of graffiti may not be enough to create an experience with a strong positive emotional valence or an overall aesthetically pleasing experience. Some participants in the Purkutaide study pointed out that in the case of graffiti, the aesthetic judgement regarding beauty is not even relevant:

I have years ago stopped assigning value to graffiti in aesthetic axis. Because they are, in a way, in some way I myself see it as an irrelevant question, such as is graffiti fine or ugly or beautiful or awful, so they are in a way secondary things, because in graffiti we play, after all, with something completely different. The dynamics in that art are born from something totally different than the aesthetic solution. (Graffiti writer, over 40 years old)

Aesthetic experience and artistic appreciation in graffiti may be related to other aspects, such as cultural knowledge and social practices. In addition, even if the work was considered visually beautiful or something else, it might still have been felt indifferently, emphasising the assumption that aesthetic judgements, emotions and art appreciation are not necessarily correlated (Gartus and Leder 2014). In general, most of the Purkutaide study participants seemed to consider the questions regarding emotions and whether graffiti was beautiful or ugly as the most difficult to answer:

One should define beautiful and ugly and so on and so forth. What word would I come up with instead of beautiful?… Beautiful is not that thing, or ugly. What is the opposite of ugly when it is not beautiful? (Graffiti writer, over 40 years old).

Verbally explicating emotions and making judgements about beauty require a conceptual definition and identification for both, and for the respondent to have an adequate vocabulary to express the finer details of the experience (Tinio and Gartus 2018). In the case of visual art or graffiti, which is produced and perceived in pictorial format, it might be very difficult or even impossible to communicate all the associated emotional experiences and inferences as spoken words, which is also a general challenge in emotion research (Barrett 2006; Frijda 2008).

6.4 Conclusions

Different perceivable and non-perceivable content seems to affect the experienced emotional episodes. The reasons why an experience is more positive or pleasant, or why specific emotions are felt, may differ from one individual in one moment to another individual or another situation, which supports the view that emotions are complex constructions of situational and subjective components. Modern appraisal theories maintain that felt emotions depend not only on the perceivable features of the assessed object, such as a work of art or graffiti writing, but also on individual-level concerns and contexts.

Visually stylistic properties or other aesthetic qualities of the graffiti work may be assessed as pleasant looking, good or even beautiful, but that does not necessarily mean that the overall experience is felt positively. Appraising graffiti involves how novel and special it seems, what kind (and how much) information a person has about the work or artist or anything else, which may affect the comprehension of the work and how interesting and engaging it is perceived to be. Appraisal also depends on the viewing time and the resources an individual has put into the evaluation process. Graffiti is notorious for being judged as ugly or 'visual litter', accompanied by hostile emotions such as anger and dislike. These kinds of emotions may depend on individual-level understanding, goals, or personal history and life experiences. Emotional appraisal also involves how the work is seemingly related to the self, such as how it matches an individual's subjective taste, standards and even their way of doing graffiti, creating emotions from pride to disappointment or causing a flat, neutral feeling. Perceiving something as 'beautiful' is itself a very complex concept. A piece of graffiti may be judged as both or neither, or rather as nice, stylish, fine or something else.

In the Purkutaide study, several participants noted that the question about what kinds of emotions the graffiti works elicit and whether they are beautiful or ugly were especially difficult to answer. This was either because it was challenging to pinpoint or name exact emotions, or because some work did not seem to elicit any emotions at all. Some respondents may have been cautious about what they said out loud to the researcher. An important question regards the methodology and methods used to study emotions. Self-reports might reveal important information, but supplementary data could be collected, for example, via videotaping, questionnaires, psychophysical measurements, or even eye tracking or brain imaging. With careful research designs and analysis, it is possible to research emotions and aesthetic experiences. What (and how) different factors influence graffiti emotions, how they can be researched, and many other intriguing and exciting questions still await answers.

References

App B, Mcintosh DN, Reed CL, Hertenstein MJ (2011) Nonverbal channel use in communication of emotion: how may depend on why. Emotion 11(3):603–617

Austin J (2010) More to see than a canvas in a white cube: for an art in the streets. City 14(1–2):33–47

Avramidis K, Tsilimpoudini M (2017) Graffiti and street art: reading, writing and representing the city. In: Avramidis K, Tsilimpoudini M (eds) Graffiti and street art: reading, writing and representing the city. Routledge, New York and London, pp 1–24

Barrett LF (2006) Are emotions natural kinds? Perspect Psychol Sci 1(1):28–58

Baumeister RF, Vohs KD, DeWall NC, Zhang L (2007) How emotion shapes behavior: feedback, anticipation, and reflection, rather than direct causation. Pers Soc Psychol Rev 11(2):167–203

Brieber D, Nadal M, Leder H, Rosenberg R (2014) Art in time and space: context modulates the relation between art experience and viewing time. PLoS ONE 9(6):e99019

Brighenti AM (2010) At the wall: Graffiti writers, urban territoriality, and the public domain. Space and Culture 13(3):315–332

Campos R (2012) Graffiti writer as superhero. Eur J Cult Stud 16(2):155–170

Carroll N (1999) Philosophy of art. A contemporary introduction. Routledge, London and New York

Clore GL, Ortony A (2008) How cognition shapes affect into emotion. In: Lewis M, Haviland-Jones JM, Barrett LF (eds) Handbook of emotions, 3rd edn. Guilford Press, New York, pp 628–642

Colombetti G (2010) Enaction, sense-making and emotion. In: Stewart J, Gapenne O, Di Paolo EA (eds) Enaction: toward a new paradigm for cognitive science. MIT Press, Cambridge, MA, pp 145–164

Cunningham WA, Zelazo PD (2007) Attitudes and evaluations: a social cognitive neuroscience perspective. Trends Cogn Sci 11(3):97–104

Dempsey A (2003) *Moderni Taide* [Modern Art]. Trans. Raija Mattila. Helsinki: Kustannusosakey- htiö Otava

Dewey J (2005) Art as experience. Perigee, New York

Dickens L (2008) 'Finders keepers': performing the street, the gallery and the spaces in between. Liminalities: A J Perform Stud, 4(1), 1–30.

Dutton D (2009) The art instinct: beauty, pleasure and human evolution. Oxford University Press, Oxford

Ekman P (1992) An argument for basic emotions. Cogn Emot 6(3–4):169–200

Ekman P (1999) Basic emotions. In: Dalgleish T, Power M (eds) Handbook of cognition and emotion. Wiley, Hoboken, NJ, pp 45–60

Fayn K, Silvia PJ (2015) States, people and contexts: three psychological challenges for the neuroscience of aesthetics. In: Huston JP et al (eds) Art, aesthetics and the brain. Oxford University Press, Oxford, pp 40–56

Ferrell J (2017) Graffiti and the dialects of the city. In: Avramidis K, Tsilimpoudini M (eds) Graffiti and street art: reading, writing and representing the city. Routledge, New York and London, pp 25–38

Fingerhut J, Prinz JJ (2018) Wonder, appreciation, and the value of art. In: Christensen JF, Gomila A (eds) Progress in brain research. Volume 237. The arts and the brain: psychology and physiology beyond pleasure (pp 107–128). Elsevier, Amsterdam

Frijda NH (2008) The psychologists' point of view. In: Lewis M, Haviland-Jones JM, Feldman Barrett L (eds) Handbook of emotions, 3rd edn (pp 68–87). Guilford Press, New York

Gartus A, Leder H (2014) The white cube of the museum versus the gray cube of the street: the role of context in aesthetic evaluations. Psychol Aesthetics, Creativity Arts 8(3):311–320

Gerger G, Leder H, Kremer A (2014) Context effects on emotional and aesthetic evaluations of artworks and IAPS pictures. Acta Physiol (Oxf) 151:174–183

Gartus A, Klemer N, Leder H (2015) The effects of visual context and individual differences on perception and evaluation of modern art and graffiti art. Acta Physiol (Oxf) 156:64–76

Gombrich EH (1963) Meditations on a hobby horse and other essays on the theory of art. Phaidon Press Ltd., London

Halsey M, Young A (2006) 'Our desires are ungovernable': writing graffiti in urban space. Theor Criminol 10(3):275–306

Iseminger G (2005) Aesthetic experience. In: Levinson J (ed) The Oxford handbook of aesthetics. Oxford: Oxford University Press. https://www-oxfordhandbooks-com.ezproxy.jyu.fi/view/10.1093/oxfordhb/9780199279456.001.0001/oxfordhb-9780199279456

Izard CE (2007) Basic emotions, natural kinds, emotion schemas, and a new paradigm. Perspect Psychol Sci 2(3):260–280

Kimvall J (2014) The G-word: virtuosity and violation, negotiating and transforming graffiti. Dokument Press, Årsta

Kuuva S (2007) Content-based approach to experiencing visual art. Dissertation, University of Jyväskylä

Leder H, Belke B, Oeberst A, Augustin D (2004) A model of aesthetic appreciation and aesthetic judgments. Br J Psychol 95:489–508

Leder H, Gerger G, Brieber D, Schwarz N (2014) What makes an art expert? Emotion and evaluation in art appreciation. Cogn Emot 28(6):1137–1147

Leder H, Nadal M (2014) Ten years of a model of aesthetic appreciation and aesthetic judgments: the aesthetic episode–Developments and challenges in empirical aesthetics. Br J Psychol 105(4):443–464

Meuleman B, Moors A, Fontaine J, Renaud O, Scherer K (2019) Interaction and threshold effects of appraisal on componential patterns of emotion: a study using cross-cultural semantic data. Emotion 19(3):425–442

Moors A, Ellsworth PC, Scherer KR, Frijda NH (2013) Appraisal theories of emotion: state of the art and future development. Emotion Revi 5(2):119–124

Myllylä MT (2018) Graffiti as a palimpsest. Street Art Urban Creativity Sci J 4(2):25–35

Panksepp J (2006) Emotional endophenotypes in evolutionary psychiatry. Prog Neuropsychopharmacol Biol Psychiatry 30:774–784

Pihko E, Virtanen A, Saarinen VM, Pannasch S, Hirvenkari L, Tossavainen T, Haapala A, Hari R (2011) Experiencing art: the influence of expertise and painting abstraction level. Front Human Neurosci 5(94):1–10

Purkutaide (2019) Purkutaide project [Online]. Available at: https://www.purkutaide.com (Accessed 20 Mar 2020)

Ross I (2016a) Introduction: sorting it all out. In: Ross JI (ed) Routledge handbook of graffiti and street art. Routledge, New York, pp 1–10

Ross I (2016b) Glossary. In: Ross JI (ed) Routledge handbook of graffiti and street art. Routledge, New York, pp 475–479

Seeley WP (2015) Art, meaning, and aesthetics: The case for a cognitive neuroscience of art. In: Huston JP et al (eds) Art, Aesthetics and the brain. Oxford University Press, Oxford, pp 19–39

Silvia PJ (2005a) Cognitive appraisals and interest in visual art: exploring an appraisal theory of aesthetic emotions. Empirical Stud Arts 23(2):119–133

Silvia PJ (2005b) Emotional responses to art: from collation and arousal to cognition and emotion. Rev General Psychol 9(4):342–357

Silvia PJ (2008) Interest: the curious emotion. Curr Direct Psychol Sci 17(1):57–60

Silvia PJ (2009) Looking past pleasure: anger, confusion, disgust, pride, surprise, and other unusual aesthetic emotions. Psychol Aesthetics, Creativity Arts 3(1):48–51

Silvia PJ (2010) Confusion and interest: The role of knowledge emotions in aesthetic experience. Psychol Aesthetics, Creativity Arts 4(2):75–80

Solso RL (2003) The psychology of art and the evolution of the conscious brain. MIT Press, Cambridge, MA

Taylor MF (2012) Addicted to the risk, recognition and respect that the graffiti lifestyle provides: towards an understanding of the reasons for graffiti engagement. Int J Mental Health Addict 10:54–68

Thompson E, Stapleton M (2009) Making sense of sense-making: reflections on enactive and extended mind theories. Topoi 28(1):23–30

Tinio PPL, Gartus A (2018). Characterizing the emotional response to art beyond pleasure: correspondence between the emotional characteristics of artworks and viewers' emotional responses. In: Christensen JF, Gomila A (eds) Progress in brain research, vol 237. The arts and the brain: psychology and physiology beyond pleasure (pp 319–342). Elsevier, Amsterdam

Tooby J, Cosmides L (2008) The evolutionary psychology of the emotions and their relationship to internal regulatory variables. In: Lewis M, Haviland-Jones JM, Barrett LF (eds) Handbook of emotions, 3rd edn. Guilford Press, New York, pp 114–137

Valjakka M (2016) Contesting transcultural trends: emerging self-identities and urban art images in Hong Kong. In: Ross JI (ed) Routledge handbook of graffiti and street art. Routledge, Abingdon and New York, pp 372–388

Van Dongen NN, Van Strien JW, Dijkstra K (2016) Implicit emotion regulation in the context of viewing artworks: ERP evidence in response to pleasant and unpleasant pictures. Brain Cogn 107:48–54

Wacławek A (2011) Graffiti and street art. Thames, Hudson Ltd., London

Young A (2005) Judging the image. Art, value, law. Routledge, Abingdon and New York

Chapter 7
Emotional Film Experience

Jose Cañas-Bajo

Abstract Technological developments in the film industry have enriched the audio–visual language over the years, and made it possible to represent subtle aspects of the world so that audiences could experience fictional stories as realistically as possible, and be emotionally engaged and interested. This chapter reviews recent research and theories regarding emotional engagement, interest and empathy as complex cognitive emotional viewers' experiences. In addition, we will discuss the use of non-narrative factors such as music or colour to elicit emotions. But more importantly, we will introduce methods and techniques that can be used to study viewers' emotional responses to films. Our basic claim is that films, like other technological artefacts, can be studied using the methods of user experiences. Integrating ideas and methods from film theory, users' experiences and cognitive psychological approaches to emotions might provide a useful framework to understand viewers' experiences of films.

7.1 Introduction

In recent decades, the rapid development of digital technology, the expansion of the Internet and the emergence of new platforms and devices have changed the film industry and made it a booming enterprise with the ability to reach audiences worldwide. Films and different types of audio–visual products are now generated to meet the diverse societal needs including entertainment, and cultural and aesthetic aims. Understanding audience reactions to these products has become critical for this purpose. This chapter reviews the theoretical approaches to cinematic emotion and recent research regarding emotional film experiences. We maintain that films,

The work in this chapter was supported by Aalto Seed and the doctoral school in the Faculty of Information Technology (University of Jyväskylä). It is based on the author's unpublished doctoral dissertation.

J. Cañas-Bajo (✉)
Aalto University, Espoo, Finland
e-mail: jose.canasbajo@aalto.fi

like other technological artefacts, can be studied by examining users' experiences with technological artefacts.

The chapter proceeds as follows. Section 2 introduces how technology has impacted the film industry and assesses some of the reactions and challenges posed by the recent digital revolution. Section 3 discusses complex cognitive emotional viewers' experiences in films—interest, engagement and empathy—which determine whether people like them. Section 4 explores the role of non-narrative factors such as music and colour in eliciting emotions. Section 5 investigates the methods and techniques that have been used to study viewers' emotions related to films, and Sect. 6 concludes.

7.2 Films and Technology

The audio–visual language has experienced many changes throughout the brief history of cinema, and continues to develop. Technological developments have made it possible to create and enrich this language: ways to change the lighting, camera movements and angles, cuts and shots, video editing, and so on have introduced new ways of communication and the possibility to represent more complex aspects of reality. The appearance of sounds and the introduction of music, voices and auditory effects to films were breakthroughs that changed approaches to interpretation. Artists continuously experiment with new techniques to produce the intended emotional effect among the spectators. However, these technological developments, and the continuous changes in how cinematographic language is used, have often been experienced with excitement and fear. Technology is sometimes viewed as a threat to the aesthetic intentions of films (Belton 2014).

For example, many critics welcomed the introduction of sound since it enhanced different ways of enriching communication. However, some filmmakers and theorists considered it to be the death of a type of cinema that mainly relied on images and visual expressions. Arnheim (1957) argued that technological advances in sound or colour were a threat to cinema as a visual art, represented by silent, black and white films. This is perceived to be even more threatening with the advent of digital images, since images of the real world are converted into numeric information for a digital processor to read. The quality of colour, light, texture and sound is registered by digital recording devices that convert the video signal into numeric information; when this information is read by digital equipment, it re-creates colours, sounds and forms to convey meaningful images.

Some analysts have argued that the digital revolution has been driven more by software and hardware companies and marketing and economics interests than by an interest in moving the movie-going experience forward (Belton 2014). After a demonstration at the 1999 Cannes film festival, the film critic Roger Ebert argued that digital movies could not duplicate the experience of analogue films, since they are too dependent on the technological device used to record the images or do the projection. Similarly, Manovich (1996) argued that digital technology has moved cinema closer to animation, and away from theatrical art. Digital technology can transform the real world into plastic objects and model and remodel them however the moviemaker wishes. While this aspect is very important for special effects in imagined, fantasy scenarios, digital projections do not offer audiences a new experience when the movie tries to reflect the real world (Belton 2002).

However, digital technologies represent interesting challenges for film theories, since they introduced completely new forms of experiencing and interpreting films (Wood 2008a). Because technology leaves traces on the screen through the emergence of competing elements (character, special effects, etc.), the viewer needs to focus their attention on different elements, which introduces choices in viewing and agency when deciding which element to attend to and include in the interpretation of the images (Wood 2008b). Digital technology introduces new lines of research and inquiries into the attentional strategies and interpretation of digital elements by the viewer, and how the audience emotionally experiences these elements.

Digital technologies have also been seen as threatening to cinema not only from the continuous changes in the style of the audio–visual language, but also from the new platforms on which films can be shown (TVs, mobile phones, tablets, etc.). They are all suitable ways of experiencing films that impose some technical constraints and affect how viewers experience the films. Since 1980, digital technologies have economically impacted the film industry through multiple synergies between hardware companies, film studios and cable companies to create integrated entertainment companies. These have resulted in 'convergence', which refers to 'the union of audio, video and data communications into a single source, received on a single device, delivered by a single connection' (Forman and John 2000, p. 50). Convergence allows the same content to be delivered from different technologies so that they increasingly resemble each other. Thus, computers are similar to televisions, tablets and mobile phones since they can be used to download and view movies from the Internet. Convergence has led to the development of video products that can be delivered via a variety of different devices. Digital technologies have enabled the studios to profit from new markets such as video, cable, television or video games, which increases the fear that the role of the theatrical release, and the immersive experience of large screens and dark silent theatres could slowly disappear (Belton 2002). Nonetheless, while these fears are still present, they do not parallel the film industry's economic growth. For example, in 2016 the European film industry increased its profits, with a record of 976.5 million cinema admissions (Talavera et al. 2016).

The use of these new platforms and convergence has also been viewed as new systems of communication that deserve theoretical and empirical studies. For example, new concepts such as remediation or irritation have emerged to reflect the changes in the environment introduced by the different platforms (irritation) that impact aesthetics and textual conventions (Wood 2008b). New communication networks have also been introduced, since convergence involves the flow between the different media, the relationship between the aesthetic elements of different platforms, PR actioners, the technological elements in the product and the viewer's interpretation of this element. While convergence has given rise to many theoretical film articles, very few empirical studies have addressed its impact on viewers' experiences. We argue that based on the integration of ideas and methods from film theory, users' experience and cognitive psychological approaches to emotions might provide a useful framework with which to address these questions.

7.3   Usability and Users' Experience

Although films are clearly designed to produce emotional experiences in the viewers, many film creators, especially in the artistic area, have focused more on generating good-quality audio–visual information than on the viewer's emotional reaction. Classic film theorists such as Rudolf Arnheim, Béla Balázs or Siegfried Kracauer (see Stam and Miller 2000 for a review) were mainly concerned with defining the crucial elements of the medium, including whether they were valid to index reality and the extent to which these elements differed from reality (Bazin 1971); they paid less attention to the target audience. Classically, professionals in the audio–visual field have been more interested in studying and developing products from the standpoint of the stimulus rather than from the viewers' perspective. However, if the audio–visual media purports to create a strong impact on viewers' minds and influence their thinking, it needs to understand their emotions, experiences and the factors influencing these experiences. Therefore, the film industry should be aware of the end users' (audience's) characteristics and put them into focus (Lang et al. 2010). Below we discuss how viewers' emotions have recently come into focus in research on cognitive and phenomenological approaches to films.

In the areas of technology and technology design, it has long been understood that users need to be at the centre of the creation process. Product design begins by understanding the users' cognitive and emotional characteristics (Garrett 2011; Helfenstein 2012; Norman 2005). The term 'usability' is a key concept: it means looking at the user to judge whether the product fulfils its functional role in a way that it is easy and pleasant to the user. We argue that for films and other video products, usability is also an important concept since they need to achieve their purpose and convey the meaning and emotions for which they were intended in an easier and more agreeable way to the viewers. In this sense, as with other technological products, aesthetic appreciation is one aspect of a product's usability. Usability includes a product's functional as well as aesthetic features—especially when it is an artistic product such as a film.

To understand the role of aesthetics in a product's perceived usability, Hassenzahl (2004, 2010) suggests that two additional constructs from the field of human–computer interactions (HCI) are of interest—perceived pragmatic quality (PQ, similar to perceived usability and ease of use) and perceived hedonic quality (related to perceived enjoyment, novelty and stimulation). Hassenzahl maintains that aesthetics plays a role in evaluating these two properties of artefacts, but in two different ways. Globally, individuals assess a product's qualities by making inferences based on the information available at the time. For example, people sometimes judge a product's worth based on its price, or its quality based on its attractiveness. This attractiveness will create a global impression of goodness that in turn will influence the product's perceived PQ. Hence, aesthetics indirectly influences PQ since it is the *impression* of global goodness that influences perceived PQ. Research on social psychology (Nisbett and Wilson 1977) has provided evidence that humans tend to assume that good-looking people also have other positive qualities. This 'halo effect' also affects products and objects: those that are perceived as attractive are also considered stimulating and enjoyable (van Schaik et al. 2012). Generally, evidence from prior studies on the relationship between usability and aesthetics shows that judgements of beauty or enjoyment are positively correlated with general goodness (Dion et al. 1972; Tractinsky et al. 2006). More importantly, beauty has a direct effect on hedonic quality, whereas its effect on PQ is indirect, mediated by goodness.

In the case of audio–visual products, the user experience or the viewer's approach can be used to address different problems (Knudsen 2002; Moggridge 2010). The viewer's approach can be applied to the devices used to display the audio–visual product and the quality of the image (Dobrian et al. 2011); it can also be employed in narrative structures, as well as technical and aesthetic aspects, also called the audio–visual language. In the case of audio–visual devices, the measures of usability have been directed to assess the viewers' experiences with the devices used to diffuse these products, and their suitability to the characteristics of such products. For example, a larger screen is needed to watch the latest Hollywood blockbuster compared to a short humorous sketch or advertisement. The physical features of a device can thus affect the viewers' experiences and engagement with the video content. However, the audio–visual experiences will be characterised by the emotional impact that the product's concrete elements provoke in the audience. Hence, this approach should also be directed to study the relationship between textual elements of the video product and how they are experienced by the viewer.

Mass communication researchers conducted pioneering studies in the twentieth century to address the media impact on people's opinions and attitudes. For instance, they studied the use of government propaganda machines to manipulate collective attitudes on a certain topic or political movement within a society (Hovland et al. 1953; Lasswell 1927). In a more academic context, audience reception studies (Hall 1980; Tomlinson 1999) have stressed the audience's central role in communication. Tyler (1992) argues that all communication is directed towards a specific audience, and that the designer of communication products needs to consider the audience as an active part of the design process. Audio–visual designers must identify the beliefs shared by the intended audience and the right references to use to persuade the audience of the credibility of an audio–visual message or the narrative sequence of a feature-length film. Audience analyses using this approach are usually performed through the use of surveys, focus groups with small representative samples of the target population or in-depth ethnographic observations of a given audience, where viewing habits are observed from the inside over a substantial period of time. Audience analyses have tried to isolate demographic factors (geographic location, race, education, gender, etc.) to explore the ways in which different groups understand and are influenced by the same message. From this perspective, the audience's system of beliefs plays a central role in the design of the audio–visual product.

Following this idea, recent research on media audiences has clearly stressed the need to understand the behaviours of the audience. Atkinson (2014) offers a rich analysis of how technology is making the audience's relationship with moving images much more interactive. These changes are reshaping the meaning of cinema by shifting the focus from the form to the audience: audience members are no longer represented by numbers and profits, but by their individual experiences as viewers. Within this framework, the important point is to analyse the audience's relationships with technology and films, and to understand viewers' emotional experiences (Livingstone and Das 2013). Audience reception studies in communication are close to the approach of users' experience design in the fields of HCI and product design. Users' analyses performed with users' design methods can be performed to identify the elements of audio–visual products that influence viewers in the desired way.

7.4 Films and Emotions

Viewers' experiences of film clearly involve emotions: 'We go to the cinema to experience mirth, compassion, sadness, bittersweet emotions, thrill, horror, and soon in response to what we see and hear happening to characters and ourselves' (Tan 2018, p. 11). Why do films offer these intense emotional experiences? What is the nature of the emotions we experience when watching a film?

In film theory, as well as in cognitive science, it is easy to identify different positions regarding the nature of emotions. Interpretive and biological positions have been discussed in cognitive science for a long time. Interpretative positions assume that emotions should be understood as judgements about a person's general state, while biological positions give much more weight to the reactions of the body (Thagard 2005). Oatley (1987) is the leading proponent of interpretative theory. He maintains that the resolution of human problems is complex as it involves multiple conflicting goals, rapidly changing environments, and rich and varied social interactions. Emotions provide an appraisal summary of the situations in which problems have to be solved, and help create focus and action (Oatley 1987). The second viewpoint emphasises bodily reactions and a more physiological approach: emotions involve brain reactions to physiological bodily changes instead of creating judgements about the overall situation. Emotions are generated through 'somatic markers' when the body sends signals to the brain (Damasio 1999). In between these two positions, some theories propose that emotions depend on bodily signals and 'cognitive appraisal' (Frijda 2013; Morrisj 2002). Thus, many researchers assume that people only experience emotions when they interpret the meaning of events occurring around them and their bodily reactions (Lazarus and Folkman 1984).

Some of the discussions regarding emotions in psychological research are also apparent in theories of emotions in audio–visual communication and film studies. The role of bodily reactions and cognitive elements in emotional processing in films and other audio–visual products has also been widely discussed (e.g., Grodal 1999). Grodal adopts an embodied theory of emotion in films based on Darwinian ideas. His position, which is close to Damasio's psychological ideas, is based on understanding films as the processes and architecture of the embodied brain. Grodal (2017, p. 4) argues:

> The brain mechanisms by which we enjoy such an embodied interaction with the world is inherited from our animal and hunter-gatherer ancestors and this explains the strong fascination with such types of narrative... The daily life of animals may also be a mise-en-scene of scenarios that represent fundamental ways of interacting with other agents in the physical world and mammals are provided with the ability to play.

Although from a very different methodological and theoretical position, phenomenological film theorists such as Marks (2000) or Sobchack (2000), in her conception of the 'lived body', argue that films involve shared affective experiences between the characters, the spectator and the sensory text of the film. These experiences often result in the 'vivid' sensations of corporeal feelings for the spectator.

This embodied concept of emotions contrasts with the positions of cognitive film theorists, which emphasise the cognitive processes behind emotional processing. Thus, cognitive film theorists (Berliner 2017; Bordwell 1989; Plantinga 2009; Silvia and Berg 2011; Smith 1995; Tarvainen et al. 2014; see Tan 2018 for a recent review) assume that viewers engage in goal-directed, unconscious processes such as inferencing and hypothesis to make sense of the film's narrative and emotions. Their views adopt some of the premises of appraisal theory: 'Emotions arise in response to the meaning structures of certain situations, different emotions arise in response to the different structures of meaning' (Frijda 1988, p. 349). According to this approach, people unconsciously assess the consequences of the events happening around them, but only experience emotions when they interpret the meanings of these phenomena. Thus, the rational assessment of a situation occurs prior to the emotional experience; therefore, the rational interpretation (appraisal) is also an element of survival (Lazarus and Folkman 1984).

Although an in-depth analysis of the positions of embodied and cognitive approaches to emotional film theory is beyond the scope of this review, we want to stress the parallel between some of the discussions in psychology, cognitive science and film theory. Similar to theories in psychology and cognitive science, most recent phenomenological and cognitive theorists assume that emotional experiences are composed of bodily reactions (affect) and cognitive processing and interpretation of these reactions. In this sense, they are also similar to the approaches coming from users' experiences, which assume that appraisal processes determine whether the user experiences they produce are pleasant or unpleasant (Saariluoma and Jokinen, 2014). From this position, cognitive film theorists have analysed viewers' emotional experiences such as engagement, interest or empathy that are central for film producers and creators, and that may determine the success of a film.

7.4.1 Viewers' Experiences: Interest, Engagement and Empathy

7.4.1.1 Interest as a Complex Cognitive Emotion

The study of emotions and emotional experiences in the arts focuses on aesthetics. One of the core emotions in aesthetics is interest (Berlyne 1974). However, traditional emotional theories have paid little attention to it (Ekman 1992; Oatley and Johnson-Laird 1996) or have considered it to be related to attention (Ortony et al. 1987) and opposite to distraction (Ortony and Turner 1990). More cognitively-oriented researchers have related interest to the tendency to acquire knowledge or certainty (Frijda 1986). Other researchers have identified interest as belonging to the family of epistemology-based emotions (Ellsworth and Scherer 2003; Rozin and Cohen 2003) and emotions associated with curiosity, exploration, and information-seeking (Fredrickson 1998; Izard et al. 2000; Tomkins 1962).

Berlyne (1960) proposed that curiosity is a strategic emotion that helps manage low levels of arousal. Stimuli with high levels of complexity, novelty and uncertainty increase arousal. Hence, curiosity (looking for novelty and complexity) helps people increase their level of arousal (Ellsworth 2013). Appraisal theories have adopted this idea to propose that interest involves: (1) appraisal of the degree to which an idea or event is complicated, unexpected and hard to process, (2) assessment of whether it is possible to cope with the difficulty, and (3) evaluation of the gains obtained from expending resources to understand (Silvia and Berg 2011; Tan 1996). Thus, this complex emotion involves appraising the novelty and complexity of the aesthetic product (e.g., film), on the one hand, and evaluating the individual's capacity and resources to cope with the complexity on the other hand (Silvia 2008).

In film theory, interest is one of the most important emotions that filmmakers and scriptwriters should generate among their audiences. Interest keeps the viewers in their seats, and motivated to expend their mental resources on the audio–visual sequences and maintain high expectations for the future. Interest gives an emotional structure to the movie and provides the motivation to keep watching it (Tan 1996). Although other emotions might be more intense and memorable, interest holds over time. The success of a film or video product depends on its ability to maintain the audience's interest for the entire duration of the video or film. According to Tan, interest is shaped by the film's narrative structure and characters, and is created and maintained by varying the level of suspense, mystery or surprise. For interest to be experienced, people need to appraise the elements of the film as novel and complex but comprehensible. Such appraisal may come from stylistics, aesthetics or the affective features of the film that may be based on combinations of multiple elements, including the types of shots, cuts, music or elements in the script. Some recent studies (Silvia and Berg 2011; Tarvainen et al. 2014) have tried to identify these elements by presenting people with short clips of pictures of different genders and asking them to rate the novelty, complexity, interest and emotions elicited by the clips. Although the participants vary in their interest ratings, diverse combinations of novelty, complexity and comprehensibility seem to predict interest.

Tan (1996) argues that the final experience with a film after watching it must be distinguished from the interest generated at each specific moment while watching it. Interest is the result of the return on the invested effort or the degree of the spectator's immersion in the film. According to Tan, interest and effort provide continuous feedback under three conditions. First, when the viewer experiences interest at a specific time, this generates an increase in the expectations for the resolution of the plot, which will later generate a higher return. Second, events that are closer to the present have more influence or weight than previous events, regardless of whether the valence is positive or negative. Third, the interest generated in a given moment will increase the effort the viewer invests in the next sequence, which in turn will enhance the interest generated in the return on the experienced effort. This continuous feedback raises the methodological point that studies that attempt to capture interest in films should continuously assess this emotion. Two studies (Cañas-Bajo et al. 2019a, b) have taken a continuous approach and asked participants to signal the sequences they experienced as interesting and central in the film in real time while they were watching feature length films (Cañas-Bajo et al. 2019a). Although these studies varied in methods and measurement approaches from those using short clips, they also showed that complex sequences eliciting different—sometimes conflicting—emotions were associated with higher levels of interest.

7.4.1.2 Engagement

Filmmakers must seek to maintain viewers' interest in the film as the sequences unfold over time, and ensure they are pleasantly engaged in the film or video product. Narrative engagement has been of interest for film theory as well as research investigating reading and narrative experiences (Bálint and Tan 2015; De Graaf et al. 2009). The experience of being 'absorbed' by a film's story or immersed in a book's narrative is common to all people who have experienced a good film or book. The experience of being 'engaged' has been defined in several ways. For example, Gerrig (2018) described the engagement experience as the sensation of travelling or being transported to the narrative world and away from reality. This experience of 'being transported' requires mental processes such as memory, attention, emotion and imagery to focus on the narrative (Green and Brock 2000).

Busselle and Bilandzic (2008) stress the multidimensional nature of engagement; they distinguish between different dimensions of attention and emotion in engagement responses to narrative text. Their studies manipulated conditions to increase/decrease engagement and measure up to 28 rating scales regarding attention, emotion and imagining experiences while reading a text. Factor analysis of the participants' responses yielded a four-factor structure—being in the narrative world, attentional focus, emotion and adopting the identity of the fictional character.

More recent accounts of engagement (Balint and Tan 2015; Tan 2018; Worth 2004) have focused on the idea of *mental simulation* in the sense that an engaged viewer is able to understand and simulate how the characters feel. As Tan (2018, 11) explains: 'Narrative runs simulations on the embodied mind just as programs run simulations on computers. I would add that film viewers take part in a playful simulation in which the film leads them to imagine they are present in a fictional world, where they witness fictional events that film characters are involved in'. Based on appraisal theory (Frijda 2007), Tan proposes that experiencing fictional events can lead to genuine emotion, since human emotions are based on interpreting and appraising the mental representation of the events. Therefore emotions can also be appraised from mental representations of imaginary and fictional events (Tan and Visch 2018).

In their empirical study, Bálint and Tan (2015) interviewed 18 individuals who reported that they were usually absorbed and engaged when watching films. In a first interview, they freely remembered their experiences of two self-selected stories. In a second interview, they were asked to retell their experiences again, but while they were visioning two self-selected segments of the two films. Their content analyses of the participants' verbal protocols were based on Johnson's (1987) proposal of image schema (or representations of embodied knowledge of one's own active experiences of the external world, including patterns and regularities), with the idea that experiences of the fictional world would include many of the schema identified by Johnson as representing the external world. Accordingly, many of the participants' verbalisations regarding absorbed narrative experiences contained several of Johnson's image schema (force, path directions, distance from body centre etc.), suggesting that absorption experiences were related to fictional simulation, where the viewer 'self travels into the centre of the narrative, the story-world it portrays. The self wants to be there, exerting force, and is taken there in some cases notably by the author. Conflicting tendencies may or may not arise. In the former case the self pushes towards the exit, while another force, possibly the author, pulls it back' (Bálint and Tan 2015, p. 20).

7.4.1.3 Empathy

Some authors, such as Stadler (2017), have emphasised the role of empathy—the ability to take the role of others and understand their behaviour—as a critical cinematic emotion. In this sense, 'empathy is the experience of the embodied mind of the other; it takes different forms and it involves a special kind of experiential understanding that can be understood as knowledge by acquaintance' (Zahavi 2014, p. 151). A film is experienced as engaging when the spectators imagine taking the place of the characters, and feel and mentally enact their experiences and emotions. Smith (1995, 2019) has proposed that empathic engagement with a character starts by recognition processes that make the viewer construct the character's mental representation, alignment processes by which spectators place themselves in relation to characters, and allegiance or evaluation processes through which spectators morally evaluate the characters.

Thus, similar to many complex emotions, empathy is assumed to have cognitive and affective components. The cognitive component involves taking a person's perspective to understand his/her feelings (Baron-Cohen 1997), which implies making inferences about the other person's mental and affective states, including thoughts, beliefs, motivations, and so on. The affective component entails being in an emotional state that is similar to that of the character the person is observing or interacting with. Some scholars propose that this component is similar to resonating with the bodily states of the observed person and is related to the functioning of mirror neurons (Gallese 2001). From a cognitive approach, empathy requires theory of mind (TOM); this requirement also applies to empathic feelings in films (Levin et al. 2013; Tan 2018). TOM refers to the cognitive representations of the intention, desires, feelings, and so on, of other people in their interactions with the world. These representations of the characters' beliefs, goals and feelings are also necessary to understand a film narrative and being engaged with the film characters (Levin et al. 2013).

Empathy and emotional engagement are central research subjects for cognitive film theorists. For example, character engagement is one of the ways in which viewers become emotionally involved. For instance, Smith (1995), Plantinga (1999) and Choi (2005) identified engagement with characters as an important source of emotions. However, film theory recognises at least two ways of understanding the term empathy: (1) the ways in which the spectator understands the character and the situation the character is in (Tan 1996) and (2) the extent to which the spectator mirror's the character's emotions (Grodal 1999; Plantinga 1999; Smith 1995). The first, often termed central imagining, is cognitive in nature, and involves imagining how it would be to stand in the character situation or in his/her state. The second type is a more automatic kind of involvement in the character's feelings or states by mirroring bodily reactions such as facial expressions and somatic reactions (Plantinga 1999; Smith 2011).

In media communication, Zillmann (1996) has also applied the notion of empathy to narrative plots in series and TV programs. According to his theory, empathy starts by observing the characters and evaluating the morality of their actions, which generates an affective disposition in favour of or against the characters. A positive affective disposition results in a person's concern about the wellbeing of the characters, sympathy for them, and taking their perspectives and roles. Negative evaluations encourage refusal and condemnation, along with a sense of counter-empathy. This empathetic or counter-empathetic feeling creates hope in the spectator that the narrative will lead to an end where the positively assessed character triumphs over the negatively evaluated one, as well as fear that this might not be so. Empathetic emotions are then important elements of communication as people differ in such feelings, and empathy can be created by different elements.

One important question regarding cinematic empathy involves finding cinematographic elements that elicit empathic emotions. Aesthetic elements such as colour, lighting or shots that focus on facial expressions are often used to help the audience relate to the characters. The next section discusses some of these cinematographic language emotional tools.

7.4.2 Cinematographic Elements of Emotional Language

As discussed above, the goal of filmmakers is to change the viewers' experiential centre so that they locate themselves in the centre of the narrative story and outside (or in the periphery of) the external (non-fictional) world. Some analysts have referred to this as a 'deictic shift' (Busselle and Bilandzic 2008; Segal 1995). Cinematographic language has some resources to induce this deictic change and encourage spectators to perceive the film narrative from the perspective of inside the fictional world. Cues for this deictic shift must differ from other narrative forms such as book reading or theatre, where the start of a chapter, the rising of the curtains or the text describing the mental states of the characters can cue the shift. Such verbal cues cannot be used in films unless a narrator is introduced to describe the characters' mental states or to provide information regarding the context to interpret the characters' actions or feelings (Busselle and Bilandzic 2008). Facial expressions, gestures and other forms of body language can convey these mental states.

As mentioned above, some theoretical approaches to emotion consider personal bodily reactions to be the building blocks of emotional experiences. However, our bodies do not only signal our *own* reactions; they also provide cues to interpreting the emotional reactions of others. In real life, emotions are not only communicated by exchanging verbal or audio–visual structured messages: the human body also plays a role in feeling and communicating emotions. We have bodily reactions to emotions; we recognise and adopt facial expressions, locate our body in different places and postures, and use gestures as communicative events (Argyle 1975).

The field of human communication and emotions studies body language. For example, Hall (1990) points out the importance of the distance between individuals when they interact and the emotions that cause people to draw closer to (or further from) the person who is speaking. Borg (2009) has shown that people can use body language to persuade and significantly influence the emotions of others. He finds that 93% of human communication depends on body language.

The roles that facial expressions, gestures and other forms of non-verbal communication play in conveying meaning and emotions is a central topic in communication research. They are critical to understand how spectators interpret bodily emotional cues from a video or film. Some theorists (Pease and Pease 2008; Walton 1984) suggest that we can determine the meaning of bodily expressions in films and media through normal observation, based on the assumption that audio-visual content is transparent (Transparency hypothesis). Transparency in this context means that observations and interpretations of facial expressions and gestures in the media are similar to observations and interpretations of body language in real life. However, many film theorists reject this position, and argue that any kind of visual representation shapes the image, and thus that the body expression represented by the image should differ (Currie 1999). The idea is that the technology used in films and audio-visual products sometimes constrains the actors' performance. For example, actors in shot-reverse-shot are often prevented from gestures and movements since they need to be closer than in real life (Naremore 1988).

In an even more extreme position, the phenomenologist film theorist Sobchack (1992) asserts that the film is an object/subject that sees and is seen. Metaphorically, the camera and the technological elements that capture the reality can be argued to constitute the 'body of the film'; this film body subjectively captures the world. When we see the film, we are subjectively experiencing the subjective view of it. According to this 'embodied' theory, bodily gestures cannot be directly observed from the film, since the film view subjectively mediates our observation. Thus, a fully transparent presentation of bodily expression in visual media is not possible. In addition, although most bodily expressions have communicative intentions, people in real communications are usually unaware of them, while such expressions are purposely enacted with clear communicative intentions in films and other audio–visual products (Hansen 2014). Hence interpretations of facial expressions and acting have become an important topic in audio–visual communication studies in general and in film in particular. It is not our aim here to review the bulk of psychological research on how facial expressions and emotions are interpreted (see Gendron et al. 2018 for a review), but to call attention to the role of facial cues in producing emotions as well as some film studies research that assesses how observing these facial and bodily cues elicits emotional embodied reactions in the spectator. The concept of emotional contagion, or vicarious experience, has been important to understanding empathic responses. Vicarious experience refers to *spontaneous sharing of feeling and perspective-taking that can be evoked by seeing, hearing about, reading about, or imaginatively simulating another person's story and situation* (Keen 2007, p. 4).

Empirical research by Raz and Hendler (2014) has shown that some cues such as dialogues, extended takes and shots including eye gaze traveling in the direction of the character's eyes elicit TOM responses, including inferences about the characters' intentions and motives. In addition, characters' gestures, bodily sounds and facial expressions have been found to induce vicarious mimicry. Guo and Staedler (in Stadler 2017) describe the results of a pilot study that used a reverse camera to record spectators' emotional responses to an emotional scene from the film *American History X*. These recorded audience responses showed strong mimicry of the characters' facial expressions, indicating that people experienced empathic emotions by imitating the bodily reactions of the characters. Other studies (Guerra 2015) show that hand-held cinematography and camera movements also elicit stronger vicarious mimicry among spectators than shots and zooms from a static camera. In general, shots and camera movements that more closely resemble the way humans move produce more empathic responses and a stronger feeling of being embedded in the film's narrative. Other formal devices such as close-ups, cuts or still frames to the character's face also enhance empathic mimicry of the character (Stadler 2017).

Lighting and colour have also been considered important tools to induce emotional reactions (Zettl 2011) and have been the subject of investigation. Prior empirical studies have demonstrated the reliability of some lighting techniques. Thus, low-key lighting characterised by a contrast between light and shadow areas and dim lights portrays sadness, fear and surprise for sad, frightening or suspense scenes, while high-key, bright lighting communicates joy. Zettl (2011) also suggests that black and white, dim lighting produces an *internal* and intimate feeling that is appropriate for some scenes, while colourful scenes make emotions external. In general, film theorists propose that colour evokes potent emotional responses in viewers, and that it is therefore an important emotional tool for filmmakers, which can also be used to shift attention to different parts of the frame. However, this research also shows that there are no universal ways of using colours, and that appropriate use of colour depends on the context and experience (Bordwell and Thompson 2004; Zettl 2011).

Because filmmakers have used soundtracks and musical scores to communicate the story's meaning and convey emotions, some studies have focused on the impact of different techniques to elicit emotional experiences in the spectators. Music is sometimes played in parallel with a given episode to enhance or diminish its emotional impact. The interplay of pitch, timing and loudness characteristics causes music to evoke different affects and moods (Levi 1982). Congruence between the mood conveyed by the music and the visual scene is assumed to intensify the emotions induced, as well as the *ironic contrast* (Bordwell and Thompson 1979; Giannetti 1982). Music that is incongruent with the visual scene or narrative has been used to elicit extreme negative effects (Bolivar et al. 1994; Boltz 2001).

7.5   Methodological Approaches

In recent decades, a considerable amount of research influenced by cognitive theories and methods has directed empirical studies that seek to understand and analyse the phenomena of film experiences. Cognitive film theory is an empirically grounded, naturalistic approach that uses methodological tools from cognitive psychology as well as conceptual approaches regarding the human mind to understand how humans experience films. According to Plantinga (2002, pp. 21–22) cognitive film theory does not necessarily imply a commitment to cognitive science, strictly defined, and certainly not to cognitive science exclusively. Cognitive film theorists tend to be committed to the study of human psychology using the methods of contemporary psychology and analytic philosophy. This can be an amalgam of cognitive, evolutionary, empirical and/or ecological psychology, with influences from neuroscience and dynamic systems theory. Thus, different cognitive theorists approach film studies using different theoretical assumptions and methods, but they share a commitment to use clear and testable assumptions to study how the human mind processes films, and demonstrate a preference for small, manageable research questions.

The complexity of the combination of the elements and processes that forms a film poses an enormous methodological problem in research on film experience (Gross and Levenson 1995; Tan 2018). Hence, cognitive film theorists seek to understand the narrative structures and cinematographic techniques that give rise to viewers' reactions to a film, which have to be broken down into smaller questions (Bordwell 1989; Cutting 2016). Thus, some studies have focused on films' structures of shots and sequences, which define different styles to characterise genres according to their meaning and/or content (Redfern 2014). Other studies have examined the psychological and perceptual aspects of the visual system's reaction to video editing, such as the cut or camera position and the movements to create impressions of continuity, disruption and motion (Bordwell 1989; Bordwell et al. 1997; Cutting and Candan 2015). Video editors try to create bridges to maintain continuity in the spatiotemporal physical features of different shots, so the viewer perceives the sequence as continuous despite the cut (Cutting 2005, 2016). Similarly, researchers have used eye-tracking technology to track the attention of experimental participants in response to changes in facial expressions and gestures, but also in response to lighting, dialogue and other film elements, emphasising the role of attentional processes as a basic aspect of film viewing (Smith 2011).

In addition, phenomenological approaches are based on subjective analyses of a small number of viewers' experiences; audience analyses are mainly based on large surveys focused on demographic analyses, and experimental studies are usually performed in very controlled environments and material looking at isolated elements. Although these studies are necessary and useful, a more holistic and naturalistic users' design approach that applies mixed methods to video and film studies might fill in the gap and be able to explore factors that impact viewers' experiences while they interact with the video in more natural contexts and with realistic materials. Future studies are needed to understand integral responses to fully scripted films, and how these emotions extend and vary through the entire sequences of events, actions and scenes composing the film. Thus quantitative and qualitative responses should be combined to encompass the wide variety of viewers' cognitive and emotional responses (Cañas-Bajo et al. 2019a).

7.6 Conclusions

Phenomenological and empirical analyses of viewers' experiences now constitute a blooming field of research that parallels and resembles the fast growth of the film and video industry. This chapter analysed some of the theories, discussion and empirical data regarding emotional responses to films. While a comprehensive discussion of the rich literature on emotional experiences to films was beyond the scope of the chapter, we introduced some of the theoretical and empirical approaches and provided an overview of previous studies on emotional viewers' experiences. We sought to draw attention to the complexity of films as multifaceted audio–visual stimuli, and to the difficulties of studying the emotional reactions of the audience to the entire sequence of events that occurs in the fictional narrative of a film. Yet, a viewer's emotional reactions may depend on their personal features (introversion–extraversion, openness to experiences, see Cañas-Bajo et al. 2019b) and cultural background (Grodal 2017). Hence further research is needed on how the emotion of entire films changes across its individual sequences, as well as how these reactions may vary for individuals with different personal characteristics or cultural backgrounds. Although it is a difficult enterprise, this challenge will enrich film theories as well as our psychological understanding of emotions.

References

Argyle M (1975) Bodily communication. International Universities Press. https://books.google.es/books?id=BbYOAAAAQAAJ

Arnheim R (1957) Film as Art: 50th anniversary printing. University of California Press

Atkinson S (2014) The performative functions of dramatic communities: conceptualizing audience engagement in transmedia fiction. Int J Commun 8:2201–2219

Bálint K, Tan ES (2015) It feels like there are hooks inside my chest: the construction of narrative absorption experiences using image schemata. Projections 9(2). https://doi.org/10.3167/proj.2015.090205

Baron-Cohen S (1997) Mindblindness: an essay on autism and theory of mind. MIT

Bazin A (1971) What is cinema? : essays selected and translated [from the French] by Hugh Gray. California University Press; /z-wcorg/

Belton J (2014) If film is dead, what is cinema? Screen 55(4):460–470. https://doi.org/10.1093/screen/hju037

Belton J (2002) Digital cinema: a false revolution. October, 98–114.

Berliner T (2017) Hollywood aesthetic: pleasure in American cinema. Oxford University Press

Berlyne DE (1960) Conflict, arousal, and curiosity. McGraw-Hill Book Company. https://doi.org/10.1037/11164-000

Berlyne DE (1974) Studies in the new experimental aesthetics: steps toward an objective psychology of aesthetic appreciation. Hemisphere Pub Corp

Bolivar VJ, Cohen AJ, Fentress JC (1994) Semantic and formal congruency in music and motion pictures: effects on the interpretation of visual action. Psychomusicol J Res Music Cogn 13(1–2):28–59. https://doi.org/10.1037/h0094102

Boltz MG (2001) Musical soundtracks as a schematic influence on the cognitive processing of filmed events. Music Percept Interdisc J 18(4):427–454

Bordwell D (1989) A case for cognitivism. Iris 9:11–40. https://qmplus.qmul.ac.uk/pluginfile.php/13140/mod_resource/content/1/david%20bordwell%20-%20a%20case%20for%20cognitivism.pdf

Bordwell D, Thompson K (1979) Film art: an Introduction. Newbery Award Records, New York

Bordwell D, Thompson K (2004) Film art: an introduction (7th ed). McGraw-Hill

Bordwell D, Thompson K, Ashton J (1997) Film art: an introduction, vol 7. McGraw-Hill, New York. https://corebutte.org/downloads/advising/A-G_Course_Outlines/a-g_Film_Studies.pdf

Borg J (2009) Body language: 7 easy lessons to master the silent language (1st ed). FT Press

Busselle R, Bilandzic H (2008) Fictionality and perceived realism in experiencing stories: a model of narrative comprehension and engagement. Commun Theory 18(2):255–280

Cañas-Bajo J, Cañas-Bajo T, Berki E, Valtanen J-P, Saariluoma P (2019) Designing a new method of studying feature-length films: an empirical study and its critical analysis. Projections 13(3):53–78

Cañas-Bajo J, Cañas-Bajo T, Berki E, Valtanen J-P, Saariluoma P (2019b) Emotional experiences of films: are they universal or culturally mediated?

Choi J (2005) Leaving it up to the imagination: POV shots and imagining from the inside. J Aesthetics Art Criticism 63(1):17–25

Currie G (1999) Cognitivism. In: Miller T, Stam R (eds) A companion to film theory. Blackwell, pp 105–122

Cutting JE (2005) Perceiving scenes in film and in the world. In: Anderson J, Anderson BF (eds) Moving image theory: ecological considerations. Southern Illinois University Press, pp 9–27

Cutting JE (2016) Narrative theory and the dynamics of popular movies. Psychon Bull Rev 23(6):1713–1743

Cutting JE, Candan A (2015) Shot durations, shot classes, and the increased pace of popular movies

Damasio AR (1999) How the brain creates the mind. Sci Am Am Ed 281:112–117

De Graaf A, Hoeken H, Sanders J, Beentjes H (2009) The role of dimensions of narrative engagement in narrative persuasion. Communications 34(4):385–405

Dion K, Berscheid E, Walster E (1972) What is beautiful is good. J Pers Soc Psychol 24(3):285

Dobrian F, Sekar V, Awan A, Stoica I, Joseph D, Ganjam A, Zhan J, Zhang H (2011) Understanding the impact of video quality on user engagement. ACM SIGCOMM Comput Commun Rev 41:362–373. https://dl.acm.org/citation.cfm?id=2018478

Ekman P (1992) Are there basic emotions? Psychol Rev 99(3):550–553. https://doi.org/10.1037/0033-295X.99.3.550

Ellsworth PC (2013) Appraisal theory: old and new questions. Emot Rev 5(2):125–131

Ellsworth PC, Scherer KR (2003) Appraisal processes in emotion. Handb Affect Sci 572:V595

Forman P, John RWS (2000) Creating convergence. Sci Am 283(5):50–56

Fredrickson BL (1998) What good are positive emotions? Rev Gen Psychol 2(3):300–319. https://doi.org/10.1037/1089-2680.2.3.300

Frijda NH (1986) The emotions. Cambridge University Press; Editions de la Maison des sciences de l'homme

Frijda NH (1988) The laws of emotion. Am Psychol 43(5):349

Frijda NH (2007) The laws of emotion. Lawrence Erlbaum Associates

Frijda NH (2013) Comment: the why, when, and how of appraisal. Emot Rev 5(2):169–170. https://doi.org/10.1177/1754073912468905

Gallese V (2001) The'shared manifold'hypothesis. From mirror neurons to empathy. J Conscious Stud 8(5–6):33–50

Garrett JJ (2011) The elements of user experience: User-centered design for the Web and beyond (2nd ed). New Riders

Gendron M, Crivelli C, Barrett LF (2018) Universality reconsidered: diversity in making meaning of facial expressions. Curr Dir Psychol Sci 27(4):211–219. https://doi.org/10.1177/0963721417746794

Gerrig R (2018) Experiencing narrative worlds. Routledge

Giannetti L (1982) Schatz,"hollywood genres: formulas, filmmaking, and the studio system"(Book Review). West Humanit Rev 36(2):176

Green MC, Brock TC (2000) The role of transportation in the persuasiveness of public narratives. J Pers Soc Psychol 79(5):701

Grodal T (1999) Moving pictures: a new theory of film genres, feelings, and cognition

Grodal T (2017) How film genres are a product of biology, evolution and culture—An embodied approach. 3:17079

Gross JJ, Levenson RW (1995) Emotion elicitation using films. Cogn Emot 9(1):87–108

Guerra M (2015) Modes of action at the movies, or re-thinking film style from the embodied perspective. Embodied Cognition and Cinema. Leuven University Press, Leuven, pp 139–154

Hall ET (1990) The hidden dimension. Anchor Books

Hall S (1980) Encoding/decoding. In: Hall S et al. (eds) Culture, media, language: Working Papers in Cultural Studies, 1972–79. Unwin Hyman, London, pp 128–139

Hansen LH (2014) The moving image: body language and media context. Kosmorama 258

Hassenzahl M (2004) The Interplay of beauty, goodness, and usability in interactive products. Hum-Comput Interact 19(4):319–349. https://doi.org/10.1207/s15327051hci1904_2

Hassenzahl M (2010) Experience design: technology for all the right reasons. Synth Lect Hum-Centered Inform 3(1):1–95. https://doi.org/10.2200/S00261ED1V01Y201003HCI008

Helfenstein S (2012) Increasingly emotional design for growingly pragmatic users? A report from Finland. Behav Inf Technol 31(2):185–204. https://doi.org/10.1080/01449291003793777

Hovland CI, Janis IL, Kelley HH (1953) Communication and persuasion; psychological studies of opinion change

Izard CE, Ackerman BP, Schoff KM, Fine SE (2000) Self-organization of discrete emotions, emotion patterns, and emotion-cognition relations. Emotion, Development, and Self-Organization: Dynamic Systems Approaches to Emotional Development, pp 15–36

Johnson M (1987) The body in the mind: the bodily basis of meaning, imagination, and reason. IL, Chicago

Keen S (2007) Empathy and the Novel. Oxford University Press on Demand

Knudsen CJ (2002) Video mediated communication (VMC)-producing a sense of presence between individuals in a shared virtual reality. In: Proceedings of international symposium on educational conferencing. https://cid.nada.kth.se/pdf/CID-206.pdf

Lang A, Ewoldsen D, Doveling K, Von Scheve C, Konijin E (2010) The measurement of positive and negative affect in media research. Handb Emotions Mass Media 79–98

Lasswell HD (1927) Propaganda technique in World War I, MIT Press

Lazarus RS, Folkman S (1984) Coping and adaptation. Handb Behav Med 282–325

Levi DS (1982) The structural determinants of melodic expressive properties. J Phenomenological Psychol 13(1):19

Levin DT, Hymel AM, Baker L (2013) Belief, desire, action, and other stuff: Theory of mind in movies

Livingstone S, Das R (2013) The end of audiences?: theoretical echoes of reception amid the uncertainties of use. In: Hartley J, Burgess J, Bruns A (eds) A companion to new media dynamics. Wiley, pp 104–121

Manovich L (1996) Cinema and digital media. Perspectives of Media Art. Ostfildern, Cantz Verlag, Germany

Marks LU (2000) The skin of the film: Intercultural cinema, embodiment, and the senses. Duke University Press

Moggridge B (2010) Designing media. MIT Press

Morrisj JS (2002) How do you feel? Trends Cogn Sci 6(8):317–319

Naremore J (1988) Acting in the cinema. University of California Press

Nisbett RE, Wilson TD (1977) The halo effect: evidence for unconscious alteration of judgments. J Pers Soc Psychol 35(4):250–256. https://doi.org/10.1037/0022-3514.35.4.250

Norman DA (2005) Emotional design: why we love (or hate) everyday things. Basic Books

Oatley K (1987) Editorial: cognitive science and the understanding of emotions. Cogn Emot 1(3):209–216. https://doi.org/10.1080/02699938708408048

Oatley K, Johnson-Laird PN (1996) The communicative theory of emotions: empirical tests, mental models, and implications for social interaction

Ortony A, Clore GL, Foss MA (1987) The referential structure of the affective lexicon. Cogn Sci 11(3):341–364

Ortony A, Turner TJ (1990) What's basic about basic emotions? Psychol Rev 97(3):315

Pease B, Pease A (2008) The definitive book of body language: the hidden meaning behind people's gestures and expressions. Bantam

Plantinga C (1999) The scene of empathy and the human face on film. In: Plantinga CR, Smith GM (eds) Passionate views: Film, cognition, and emotion. Johns Hopkins University Press, pp 239–255

Plantinga C (2002) Cognitive film theory: an insider's appraisal. Cinémas: Revue d'études cinématographiques 12(2):15. https://doi.org/10.7202/024878ar

Plantinga C (2009) Moving viewers: American film and the spectator's experience. University of California Press

Raz G, Hendler T (2014) Forking cinematic paths to the self: Neurocinematically informed model of empathy in motion pictures

Redfern N (2014) Quantitative methods and the study of film. https://www.academia.edu/download/33712649/Nick_Redfern_-_Quantitative_methods_and_the_study_of_film.pdf

Rozin P, Cohen AB (2003) Reply to commentaries: confusion infusions, suggestives, correctives, and other medicines. Emotion 3(1):92–96. https://doi.org/10.1037/1528-3542.3.1.92

Saariluoma P, Jokinen JPP (2014) Emotional dimensions of user experience: a user psychological analysis. Int J Hum-Comput Interact 30(4):303–320. https://doi.org/10.1080/10447318.2013.858460

Segal EM (1995) Narrative comprehension and the role of deictic shift theory. Deixis Narrative: Cogn Sci Perspect 3–17

Silvia PJ, Berg C (2011) Finding movies interesting: how appraisals and expertise influence the aesthetic experience of film. Empirical Stud Arts 29(1):73–88. https://doi.org/10.2190/EM.29.1.e

Smith M (1995) Engaging characters: fiction, emotion, and the cinema. Oxford University Press, Clarendon Press

Smith M (2011) Empathy, expansionism, and the extended mind. Empathy: Philos Psychol Perspect 1

Smith M (2019) Film, art, and the third culture: a Précis

Sobchack VC (1992) The address of the eye: a phenomenology of film experience. Princeton University Press

Stadler J (2017) Empathy in film. Routledge Handb Philos Empathy

Stam R, Miller T (eds) (2000) Film and theory: an anthology. Blackwell

Talavera J, Kanzler M, Fontaine G (2016) European audiovisual observatory (2016). In: Public financing for film and audiovisual content–the state of soft money in Europe. European Audiovisual Observatory. Public Funding for Film and Audio-Visual Works in Europe: Key Industry, Strasbourg, p 173

Tan ES (1996) Emotion and the structure of narrative film: Film as an emotion machine. Erlbaum

Tan ES (2018) A psychology of the film. Palgrave Commun 4(1). https://doi.org/10.1057/s41599-018-0111-y

Tan ES-H, Visch V (2018) Co-imagination of fictional worlds in film viewing. Rev Gen Psychol 22(2):230–244

Tarvainen J, Sjoberg M, Westman S, Laaksonen J, Oittinen P (2014) Content-based prediction of movie style, aesthetics, and affect: data set and baseline experiments. IEEE Trans Multimedia 16(8):2085–2098. https://doi.org/10.1109/TMM.2014.2357688

Thagard P (2005) Mind: introduction to cognitive science (2nd ed). MIT Press

Tomkins SS (1962) Affect, imagery, consciousness. Springer Pub Co

Tomlinson J (1999) Globalization and culture. University of Chicago Press

Tractinsky N, Cokhavi A, Kirschenbaum M, Sharfi T (2006) Evaluating the consistency of immediate aesthetic perceptions of web pages. Int J Hum Comput Stud 64(11):1071–1083. https://doi.org/10.1016/j.ijhcs.2006.06.009

Tyler AC (1992) Shaping belief: the role of audience in visual communication. Des Issues 9(1):21. https://doi.org/10.2307/1511596

van Schaik P, Hassenzahl M, Ling J (2012) User-experience from an inference perspective. ACM Trans Comput-Hum Interac 19(2):1–25. https://doi.org/10.1145/2240156.2240159

Walton KL (1984) Transparent pictures: on the nature of photographic realism. Crit Inq 11(2):246–277. https://doi.org/10.1086/448287

Wood A (2008a) Cinema as technology: encounters with an interface

Wood A (2008) Encounters at the interface: distributed attention and digital embodiments. Q Rev Film Video 25(3):219–229. https://doi.org/10.1080/10509200601091490

Zahavi D (2014) Self and other: exploring subjectivity, empathy, and shame. Oxford University Press, USA

Zettl H (2011) Sight, sound, motion: Applied media aesthetics (6th ed). Wadsworth Cengage Learning

Zillmann D (1996) Sequential dependencies in emotional experience and behavior. Emot Interdisc Perspect 243–272

Chapter 8
The Role of Cuteness Aesthetics in Interaction

Stuart Medley, Bieke Zaman, and Paul Haimes

Abstract While many modern cultures around the world appreciate 'cuteness', few empirical studies have been conducted on the kinds of responses cuteness evokes. This chapter explores the results of two studies to examine people's perceptions and preferences regarding cute aesthetics. The first study investigated 2D online gambling aesthetics in video games and compared cute versus non-cute imagery of a croupier and a treasure chest. A total of 37 adults participated in this online experiment, which featured open and closed question items. The adult participants ($n = 17$) who took part in the second study were shown a video of a 3D ambient media device, called Fuji-chan, designed to provide information about the meteorological conditions on, and the volcanic activity of, Mount Fuji in Japan. Participants were then invited to answer questions related to the perceived cuteness, information usefulness and importance of the Fuji-chan device. The findings of both studies show that an aesthetic design that follows the principles of cuteness does not guarantee that the imagery is perceived as such, and that the content of the imagery determines whether people evaluated cuteness at a sensory, aesthetic level or whether they attached a symbolic, situated meaning to it. We call on future work to elaborate a clear operationalisation of what constitutes cuteness, at both linguistic operational and aesthetic levels, and further this preliminary work on how people's perceptions of and responses to cuteness in interaction depend on the context.

S. Medley (✉)
Edith Cowan University, Perth, Australia
e-mail: s.medley@ecu.edu.au

B. Zaman
KU Leuven, Leuven, Belgium
e-mail: bieke.zaman@kuleuven.be

P. Haimes
Ritsumeikan University, Kyoto, Japan
e-mail: haimes@fc.ritsumei.ac.jp

© Springer Nature Switzerland AG 2020
R. Rousi et al. (eds.), *Emotions in Technology Design: From Experience to Ethics*, Human–Computer Interaction Series,
https://doi.org/10.1007/978-3-030-53483-7_8

8.1 Introduction

Many modern cultures appreciate cuteness (Marcus et al. 2017). Yet the perception of cuteness lacks the clarity of terms used to describe the perception of other emotions, such as 'scary' to describe fear (Buckley 2016). This lack of clarity may contribute to difficulties in operationalising empirical measures for cuteness.

An appreciation of cuteness clearly plays a social function. Beyond simple aesthetic choices, cuteness elicits a caring, parental response, and may evoke feelings of empathy and compassion (Kringelbach et al. 2016). 'Cute emotion' is thus recognised and used widely and pervasively in modern human societies (Marcus et al. 2017), despite having no specific name, other than vernacular expressions such as 'aww' in English. There are various possible reasons for this. The most plausible is that only modern civilisations have had the luxury of recognising and responding to cuteness through the deliberate design, manufacture, marketing, and sale of items perceived as cute and used in social interactions (Buckley 2016).

In the context of interaction design, in addition to shaping emotions, artefacts perceived as cute are often employed to convey information to the user. Previous research has noted that cute artefacts can focus the user's attention (Nittono et al. 2012). However, our chapter may problematise the functions of cuteness in conveying information and focusing attention.

This chapter outlines some definitions of cuteness, particularly the accepted visual aspects, and explores previous research on how this visual aesthetic may be achieved. We find that there is some confusion in the literature regarding what constitutes 'cuteness', and use this finding to build a problem statement. We then discuss the results of a user study on visualisations within an online gambling system, as well as early user trials of a system for conveying meteorological and hazard information related to Japan's Mount Fuji. In our analysis, we discuss how the study participants perceived cuteness, and how this is related to interacting with a designed artefact. In both cases, the artefacts' users appear to have questioned the emotion conveyed by the cute aesthetics we have employed, suggesting that context strongly affects users' responses. Rather than eliciting an endearing response such as 'aww', the participant responses in our studies suggest that the aesthetics in these interactive contexts have produced cognitive dissonance among users.

8.1.1 The Aesthetics of Cuteness: Some Definitions

Marcus et al. (2017, 93) define cuteness as 'a characteristic of a product, person, thing, or context that makes it appealing, charming, funny, desirable, often endearing, memorable, and/or (usually) nonthreatening'. In post-war Japanese culture, 'cute' is referred to as kawaii (可愛い) in Japanese, a term that is derived from the Chinese word keiai (可爱), which also means 'cute' (Chen 2014). Since the end of the Second World War, Japan's subservience and deference to the United States has manifested

しばらく
お待ち ください
Please Wait
for a moment

Fig. 8.1 Pipo-kun, the Tokyo police department's mascot, featured on a sign near Meiji-Jingu shrine in Tokyo. Photo by Paul Haimes

itself in a multitude of ways across a popular culture that values childishness and cuteness (Sato 2009). This symbolic trend of perceived pacifism is widely visible in Japan, where cute characters are frequently used to confer information to the public, provide warnings and even represent police authority (see Fig. 8.1).

Previous research (Lorenz 1971; Ngai 2005, 816, 827; Marcus et al. 2017, 64–65) has suggested that artefacts with rounded and soft aesthetics are associated with cuteness, and that body proportions have an effect on perceived cuteness (Alley 1983, 621; Cho 2012). Caricatured features, such as unnaturally big heads or eyes, are often used to make a character or artefact appear cute (Weeks 2018, 131–132). Such features evoke a motivation to care or empathise by conjuring a form that resembles an infant human or pet. This was articulated in the concept of baby schema or Kindchenschema, first proposed by Lorenz (1971).

Distorting features through drawing has become a convention of Japanese illustration culture. Manga and anime often employ the 'super-deformed' style. Entire cartoons may be made in this way, or visually realistic animations or comics may momentarily switch to this mode. Exaggerating a character's features in this way—typically by enlarging their head—obscures their identity in order to amplify the emotion of a moment in the narrative.

Ohkura et al. (2013) have explored the cuteness of tactile materials, noting that yarn, cotton and sheep pile fabric were considered the most 'kawaii', while coarse materials such as sand or granite were not considered kawaii. These researchers highlighted that the Japanese onomatopoeia (a word that both denotes an object or action, but also the sound made by the object or action) describing the tactile experiences of these materials also reflected whether they were kawaii or not. These findings were consistent across genders and all age groups. Onomatopoeia that began with a harder-sounding consonant (e.g., similar to the English consonants of j, z, g) were generally used to describe the less kawaii objects, while the more kawaii objects were described with softer consonants (such as p and m). Japanese onomatopoeia has been used to demonstrate that sound-symbolic words connoting sensory information can be easily taught to non-Japanese speakers, suggesting that their descriptive sounds of cute aspects may cross language boundaries. This distinction between harder and softer sounds and objects is also found with the bouba/kiki effect (Ramachandran and Hubbard 2001), where the coarser, sharper object (kiki) is considered, across language groups, to be coarser than the softer, rounder object (bouba).

In addition to this wide variety of sensory modes, Nenkov and Scott (2014, 327) observe: 'cute products (e.g., an ice-cream scoop shaped like a miniature person or a dress with tropical colours and pink flamingos) can also have a whimsical nature, which is associated with capricious humour and a playful disposition'. Unlike the application of Kindchenschema to elicit protective feelings or behaviours towards a defenceless creature, whimsical cuteness derives from playfulness perceived in the artefact.

8.1.2 Problem Statement

In summary, we found that there is some ambiguity in the literature regarding what constitutes 'cuteness'. This confusion may be compounded by the absence of a clear term, in any culture or language, to communicate the emotional response to a beholder's perception of cuteness. Likewise, the literature contains several, quite different ways of achieving a cute aesthetic, including through the application of Kindchenschema—where the methods are relatively clear—or the more vague application of 'whimsical' or 'playful' aesthetics. Furthermore, as Nenkov and Scott (2014) explore in *'So Cute I Could Eat It Up': Priming Effects of Cute Products on Indulgent Consumption*, the effects of perceiving cute products on people's attitudes and behaviour (e.g., caring, self-control) are not unambiguous. To address this ambiguity, we present two studies—one based on two-dimensional software graphics and another based on a 3D-printed interface—that rely on Kindchenschema and rounded visual elements and smooth textures to convey a sense of cuteness to the user. By empirically measuring and contextualising people's responses to these different aesthetics of cuteness, we aim to better understand how cuteness 'works', that is how people perceive it and how it may trigger certain cognitive and emotional reactions in particular contexts.

8.2   Study of Visually Cute Aesthetics in a Simulated Online Gambling Environment

8.2.1   Goal of the Study

The first of the two studies explored in this chapter investigates the effects of cuteness aesthetics on users in an online gambling environment. It seeks to better understand the role of aesthetics and its contribution to debates on gaming literacy. This project is an interdisciplinary response to what the researchers observe in the convergence between gaming and gambling aesthetics, in order to develop strategies that improve sensitivity to persuasive messages.

The authors sought to determine the role of cuteness vs. non-cute aesthetics in the game environment in shaping gambling attitudes and aesthetic preferences. We were particularly interested in the potential paradox between (a) cuteness aesthetics eliciting a more caring approach to a task (as documented in Sherman et al. 2009) and (b) cuteness aesthetics eliciting a fondness towards an artefact that might relax the user's defensive attitude, and perhaps cause financial harm.

We ran several experiments to test various hypotheses regarding the influence of game aesthetics on players' attitudes and behaviours towards gambling. We tested whether cute aesthetic cues might lead to more or less indulgent gambling behaviour and more or less relaxed attitudes towards gambling compared to non-cute alternative conditions. Here we present only the manipulation check for the aesthetics portion of this series of experiments.

8.3   Participants

Anonymous participants ($n = 43$) aged 18 and over were recruited via social media to respond to a Qualtrics survey; 37 completed the survey (23 males, 14 females). Each respondent confirmed that they were over 18 years of age. Participants' ages ranged from 19 to 73 (mean 35 years, SD = 13).

8.4   Measures

The survey consisted of seven main building blocks: (1) informed consent, (2) demographics and background, (3) questions on the aesthetics of the treasure chest, (4) questions on the aesthetics of the croupier imagery, (5) comparison treasure chest aesthetics, (6) comparison croupier aesthetics, (6) gaming and gambling engagement, and (7) debriefing and prize draw.

Two types of aesthetics represented the image of the treasure chest (Fig. 8.2) that respondents had to consider in order to answer the question items in blocks 3 and 5

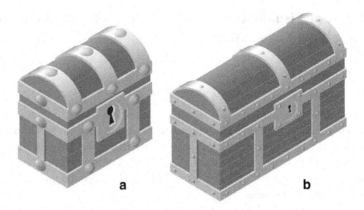

Fig. 8.2 Cute (left) and non-cute (right) conditions of the treasure chest

of the survey, and two versions of the croupier image (Fig. 8.3) that was the subject of blocks 4 and 6. Except for blocks 5 and 6, where respondents were shown both versions to elicit their preference, they were assigned to one condition only for all previous blocks.

The illustrations were designed to accord with directions implicit in the literature on cuteness. The treasure chest was illustrated (a) with proportions exaggerated to evoke 'cuteness' (cute condition) and (b) with realistic proportions (non-cute condition). The croupier was designed (a) with exaggerated child-like features (large eyes relative to the face, large head out of proportion with the body) to evoke 'cuteness' (cute condition) and (b) realistically (non-cute condition) (Lorenz 19,471 Alley 1981; Little 2012).

While the researchers developed two conditions (one cute and one non-cute) for each pair of subjects (treasure chest and croupier), participants were randomly assigned only one condition to evaluate, accompanied by the following question: 'We are interested in your opinion about an illustration and invite you to evaluate it.' We first asked open questions and encouraged them to, 'Please indicate the top five

Fig. 8.3 Cute (left) and non-cute (right) conditions of a gambling croupier

words that come to mind when seeing the illustration.' This prompt aimed to avoid priming respondents with any terms that may elicit responses about cuteness. We then asked a series of closed questions (e.g., rating the perceived level of cuteness on a seven-point scale from strongly disagree to strongly agree with the statement 'the illustration is cute'). Finally, each respondent was shown both versions to compare and indicate their preferred one and asked to explain why. Open questions were analysed qualitatively. Closed questions were analysed quantitatively using SPSS software.

8.5 Results

8.5.1 Online Gambling Experiment

8.5.1.1 Treasure Chest

Each respondent was randomly shown one version of the treasure chest and asked to 'indicate the top five things that come to mind when seeing the illustration'. No respondents included 'cute' in their five words. The closest in meaning to 'cute' were words such as 'small', 'toy' and 'casual'. Other word choices alluded to the contexts in which one might expect to find the illustrations: 'video game', 'cartoon', 'mobile game'.

On average, respondents who were asked to evaluate the cute treasure chest illustration ($M = 4.42$, SE $= 0.32$) did not evaluate this image as significantly cuter than those who were shown the non-cute image ($M = 3.56$, SE $= 0.40$). Hence, the difference in perceived cuteness was not significant: $t(35) = 1.71, p > 0.05$.

At the end of the survey, participants were encouraged to compare both versions of the treasure chest and indicate which one they preferred. This was the first time they saw the other version. Nearly three-quarters of respondents (27, 73%) selected the cute treasure chest; five respondents (13.5%) selected the non-cute condition and five respondents left this question blank (13.5%).

Participants were then prompted to explain the motivations behind their responses. Of the 33 motivational statements submitted, six of the responses that preferred the cute condition used the word 'cute' or 'cuter'. Others alluded to aspects of cuteness as detailed in Marcus et al. (2017, 93), such as 'appealing' and 'childlike'. The remainder mentioned the context in which cute aesthetics are often found, such as computer games or cartoons (as per the responses to the 'top-of-mind' question at the start of the questionnaire). The responses that indicated the cute condition as the preference are as follows:

> It's cuter in a sense. More appealing.
>
> [A, i.e. the cute version] is smaller and easier to see, it looks more like a cartoon illustration that I would imagine.
>
> It's more cartoon-like, plus would fit better on a portrait aspect ratio mobile phone ;)

It's cute, not overburdening.

It's more visually appealing. The box on the left is something I'd expect to see in a casual mobile game. The box on the right is something I'd expect to see in an isometric game aimed at a core gaming audience. I wouldn't expect either to be used as a lootbox—more as an in-game container containing items.

There's a level of abstraction that I like. Plus it reminds of the treasure chests in Zelda (one of my favourite games).

More 'cartoon', therefore 'friendlier' in appearance—less thought provoking, i.e. less of a threat.

It looks cuter and more childlike due to the larger graphic elements. If the idea is to provide an image that is not serious then the one on the left does this better than the one on the right.

It looks cute, more cartoony, like you would see in a children's pirate cartoon show.

Looks more childlike.

Tighter design, more cartoony.

I think the bigger elements (bigger keyhole, more 'gold') and the smaller overall size make it... cuter, more appealing, more fun maybe. It's more cartoony.

8.5.2 Croupier

The top-of-mind associations linked to the two croupier illustrations yielded similar results: 'gambling' and 'casino' were the most popular words in both conditions, and cuteness did not feature in the top-of-mind associations for either croupier picture. However, there were subtle differences in the top-of-mind associations between the two conditions: one respondent in the cute condition mentioned 'cartoon' and another suggested 'young', and one respondent in the non-cute condition made a reference to 'ugly'.

Respondents who saw and rated the cute croupier on cuteness ($M = 4.71$, SE $= 0.31$) did evaluate this image significantly cuter than those who were shown the non-cute croupier image ($M = 2.47$, SE $= 0.36$). This difference was significant $t(30) = 4.66, p < 0.01$. Also in contrast to the responses regarding the treasure chest illustrations, when participants were asked to indicate their preferences regarding the croupier illustrations, the choice was much more evenly split: 17 respondents (46%) chose the cute image and 15 (40.5%) the none-cute version; 5 respondents did not answer this question (13.5%).

Participants were again prompted to explain the motivations behind their preferences. At least one participant reported preferring the cute illustration of the croupier, citing the same reason as for their preference in the treasure chest choice ('Same reason as with the loot box. Seems more innocent.'). However, the majority of the responses suggested that the participants were somewhat disturbed by a croupier illustration employing child-schema techniques in the adult context of gambling. Reasons given for preferring the non-cute version included:

At least I am playing with an adult.

If this illustration is to be used in a casino themed/gambling game, the more realistic version would be better suited

Closer to the appearance of a human, more fitting for a mature game about gambling

It looks more realistic and less sinister. The one on the [left, i.e. cute version] is too childlike which makes me feel uncomfortable as children are not typically connected with casinos and gambling environments

I prefer [B, i.e. the non-cute version] over [A, i.e. the cute version], as the woman looks old enough to be a croupier.

The [second, i.e. the non-cute version] image establishes a (marginally) greater sense of trustworthiness and professionalism.

8.5.3 Discussion of the Findings

The survey highlights an empirical perspective on the perception and preferences of imagery that is designed to be more or less cute. The findings reveal that an aesthetic design that follows the principles of cuteness does not guarantee that the imagery is perceived as such. Our manipulation was successful for the croupier image but not for the treasure chest; respondents preferred the non-cute version of the croupier, in part because it appears to be more age appropriate in the gambling context. This seems to demonstrate that people can judge an image beyond the aesthetic, sensory level and attach a cultural, situated meaning to it. The treasure chest did not seem to evoke this cultural level: respondents' preferences were more closely related to their evaluation of the image's aesthetic qualities.

8.6 Fuji-Chan Device Experiment

Fuji-chan is an ambient media device designed to provide information about the meteorological conditions on, and the volcanic activity of, Mount Fuji in Japan (Fig. 8.4). Although it is one of the country's best-known sights, few people are aware that Mount Fuji is an active volcano (JMA 2018). Although it has not erupted in over 300 years, the Japanese government predicts that an eruption would require widespread evacuations of several nearby towns and prefectures, and debris and ash could reach as far as Tokyo (Aramaki 2007).

The exterior was created with a 3D printer and was inspired by abstract, comical representations of Mount Fuji (e.g., Fig. 8.5). The interior contains LED lights, a piezo buzzer and a Wi-Fi module. The device is powered by a standard mini-USB connection. Photo by Paul Haimes.

The Fuji-chan device uses 'internet of things' technology to connect to a remote server that uses weather data sourced from myweather2.com. It also retrieves the current eruption warning level from the National Institute of Informatics, which uses data from the Japan Meteorological Agency. The peak of the device uses a programmable multi-colour LED (light emitting diode) light to convey weather

Fig. 8.4 Second prototype of the Fuji-chan device

information (warmer and cooler colours represent, respectively, warmer and cooler temperatures), while the base colour corresponds to the level of volcanic eruption warning (Table 8.1). If the warning level is three (which requires evacuation) or above, the device will flash and an alarm will sound.

The device was designed to be cute. Its exterior features rounded, soft edges and is based on comical representations of Mount Fuji such as that shown in Fig. 8.5, which bear little resemblance to the real volcano. Indeed, its name is a play on Japanese suffixes to emphasise its cuteness. Mount Fuji is referred to as 'Fuji-san' in Japanese; 'san' refers to a mountain but is also a standard suffix used when addressing adults regardless of gender. 'Chan' is a suffix usually applied to those who the speaker feels a great deal of affection for, such as young children, pets and romantic partners (Mogi 2002). The cute, playful nature of Fuji-chan's exterior and name may seem at odds with an object created to warn of possible disaster, but as explained earlier (e.g., Fig. 8.1), there is a tendency in Japan to use cuteness in contexts of authority.

The Fuji-chan device has been demonstrated at multiple exhibitions throughout Asia. Without being prompted, several observers at these exhibitions have noted that the device had a cute appearance. Two noted that the name was 'kawaii' (cute). An unexpected outcome of these demonstrations was that no attendees reported that the device was useful or provided important information. This outcome highlighted the need to conduct further user testing of the device.

A testing session was thus conducted with 17 users (10 females, 7 males) over the age of 18. For this user study, participants watched a short video explaining how the Fuji-chan device works and could also see it being demonstrated in a scenario with both live data (i.e., near real-time conditions) from the server and testing data to demonstrate the device's behaviour during a major (Level 5) eruption event. The live data showed a temperature of 17 degrees Celsius and an eruption warning of Level 1. Afterwards, participants were asked three questions, as described in Table 8.2.

Fig. 8.5 A comical representation of Mount Fuji on a Fujisan Express train between Otsuki and Kawaguchiko in Yamanashi prefecture, Japan Photo by Paul Haimes

Answers were provided in the form of a Likert scale, where 1 = Strongly disagree, 2 = Slightly disagree, 3 = Neither agree nor disagree, 4 = Slightly agree and 5 = Strongly agree. All participants were current residents of Japan, both Japanese nationals ($n = 8$) and foreign residents ($n = 8$), with the exception of one visitor to Japan who had visited within a month before the study.

Overall, participants strongly agreed that the Fuji-chan device was cute (Question 1), with little deviation in participant answers. Responses were less positive to the question (2) regarding the importance of the information provided. For example,

Table 8.1 Japan Meteorological Agency volcanic eruption warnings

Level	Colour	Meaning
1	White	Potential for increased activity
2	Yellow	Do not approach the crater
3	Orange	Do not approach the volcano
4	Red	Prepare to evacuate
5	Purple	Evacuate

Source JMA (2018)
Note the Fuji-chan device uses the same colour warning system for its volcanic eruption warnings

Table 8.2 Results of user testing ($n = 17$)

Question	Mean	SD	Positive
1. Do you think the Fuji-chan device's appearance is cute?	4.94	0.24	17/17
2. Do you think the information provided by Fuji-chan is important?	3.47	0.70	10/17
3. Does the Fuji-chan device seem useful?	3.76	1.11	12/17

Note This table shows the three questions asked, the mean (average) answers, the standard deviation (SD) and the number of positive responses (4 or 5 on the five-point Likert scale)

one participant, a first-time visitor to Japan, was unaware that Mount Fuji was an active volcano. Responses regarding the usefulness of the device (Question 3) showed the greatest deviation, but were overall slightly more positive than the answers to Question 2. Due to a lack of strong positive or negative correlation, it is unclear whether cuteness has a positive or negative impact on users' perceptions of the importance of the information being conveyed and perceived usefulness, but based on responses from the exhibition attendees and the data in Table 8.2, it is possible that cute aesthetics could cause users to overlook the importance of the information being presented to them.

8.7 Discussion and Conclusion

The online gambling study, the deployment of two sets of illustrations (treasure chest and croupier), with cute and non-cute versions of each, demonstrated that what is depicted as cute affects participants' preferences. The survey questions did not clearly elicit words from respondents that could be construed as meaning 'cute' about any of the illustrations when they were viewed in isolation. However, when the pairs of illustrations were shown together and respondents were asked to explain their preferences, the findings suggest that the content of the image determines whether respondents evaluate it at a sensory level or a more symbolic level. The manipulation of the aesthetics of the treasure chest was not successful, as several respondents did not rate the design that was meant to be cute as such. However, most respondents clearly preferred the cute version. While there was a definite consensus on which croupier image was the 'cute' illustration, preferences were evenly divided between the cute and non-cute versions. While Marcus et al. (2017, 93) define cuteness as 'a characteristic of a product, person, thing, or context that makes it appealing, charming, funny, desirable, often endearing, memorable, and/or (usually) nonthreatening', some participants observed that one of the croupier images was indeed cute, but did not find it 'appealing', 'charming' or 'endearing' due to the perceived inappropriateness of cuteness in the context of gambling.

Based on the results from the exhibition and user research of the Fuji-chan device, we suggest that when designers create an artefact to elicit an appreciation of cuteness from the user, it is worth considering whether this response may inadvertently detract from the device's intended purpose. That is, the aesthetic level should not negatively influence the cognitive processing required for the user to process the information provided. Further studies are required to determine whether cuteness can be a distracting factor when conveying meaning in an interaction. After all, underlying Japan's facade of modern culture steeped in cuteness and childishness belies a stagnating economy and society. Rather than eliciting a response of 'aww', cute designs in inappropriate contexts appear to have generated cognitive dissonance.

The study suffers from at least three limitations. First, the aesthetics we presented as part of our research instruments were removed from their natural settings, which threatens the ecological validity of our findings. Other elements of the digital and—in the case of Fuji-chan—physical environment where our artefacts would normally be found may affect the user's aesthetic response.

Second, our survey participants were instructed to comment on the visual conditions developed for our manipulation checks. Palmer et al. (2012) suggest that everyone has an aesthetic response to everything they see, but they may not be conscious of this response except under extraordinary conditions, such as when their attention is directed to respond because of the context. Our directing the participants may have affected, if not their response, at least the intensity of that response.

Finally, we acknowledge that while we have striven to design our test artefacts according to the principles of cuteness, these principles are not consistently defined in the literature. Overcoming the relative vagueness regarding the principles of cute aesthetics is a more protracted and complex problem. Developing a clear operationalisation of what constitutes cuteness, at both the linguistic and aesthetic levels, will advance experiments in cuteness response, but will take time and a concerted effort. We hope our work has helped to highlight that this effort should consider how cognitive, symbolic and contextual parameters co-shape the aesthetic experience.

References

Alley TR (1981) Head shape and the perception of cuteness. Dev Psychol 17(5):650–654
Alley TR (1983) Age-related changes in body proportions, body size, and perceived cuteness. Perception and Motion Skills 56:615–622

Aramaki S (2007) Volcanic disaster mitigation maps of Fuji Volcano and overview of mitigation programs. Yamanashi Institute of Environmental Sciences. https://www.mfri.pref.yamanashi.jp/fujikazan/web/P451-475.pdf.

Buckley RC (2016) Aww: the emotion of perceiving cuteness. Frontiers in Psychology 7:1740

Chen H (2014) A study of Japanese loanwords in Chinese. Master's thesis, University of Oslo

Cho S (2012) Aesthetic and value judgment of Neotenous objects: cuteness as a design factor and its effects on product evaluation. Dissertation, University of Michigan

Japan Meteorological Agency (2018) Volcanic warnings. https://www.jma.go.jp/en/volcano/.

Kringelbach ML, Stark EA, Alexander C, Bornstein MH, Stein A (2016) On cuteness: unlocking the parental brain and beyond. Trends Cognitive Sci 20(7):545–558

Little AC (2012) Manipulation of infant-like traits affects perceived cuteness of infant, adult and cat faces. Ethology 118(8):775–782

Lorenz K (1971) Studies in animal and human behavior. Harvard University Press, Cambridge, MA

Marcus A, Kurosu M, Ma X, Hashizume A (2017) Taxonomy of cuteness. In: Marcus A et al (eds) Cuteness Engineering. Springer, Cham, pp 93–118

Mogi N (2002) Japanese ways of addressing people. Investigationes Linguisticae 8:14–22

Nenkov GY, Scott ML (2014) 'So cute I could eat it up': Priming effects of cute products on indulgent consumption. J Consumer Res 41(2):326–341

Ngai S (2005) The cuteness of the avant-garde. Critical Inquiry 31(4):811–847

Nittono H, Fukushima M, Yano A, Moriya H (2012) The power of kawaii: viewing cute images promotes a careful behavior and narrows attentional focus. PLoS ONE 7:e46362

Ohkura M, Osawa S, Komatsu T (2013) Kawaii feeling in tactile material perception. In: Proceedings of the 5th international congress of international association of societies of design research, Tokyo.

Palmer SE, Schloss KB, Sammartino J (2012) Hidden knowledge in aesthetic judgments: preference for color and spatial composition. In: Shimamura A, Palmer SE (eds) Aesthetic science: Connecting minds, brains and experience. Oxford University Press, Oxford, UK, pp 189–222

Ramachandran VS, Hubbard EM (2001) Synaesthesia—a window into perception, thought and language. J Consciousness Stud 8(12):3–34

Sato K (2009) A postwar cultural history of cuteness in Japan. Education About Asia 14(2):38–42

Sherman GD, Haidt J, Coan JA (2009) Viewing cute images increases behavioral carefulness. Emotion 9(2):282–286

Weeks KK (2018) Cuteness and appeals: unleashing prosocial emotions. Dissertation, University of Connecticut.

Part IV
Ethics and Culture

Cultures have their rules. These rules are determined by the consequences of actions and endpoints of respective social discussions, interactions and relationships. Cultural rules, social and moral codes, and what is defined as being 'the way to be' in the world as well as in relation to other people, play crucial role in the analysis of emotions. That is, when establishing measures to systematically ascertain what emotional qualities will be experienced in relation to what design element or property, one of the key factors to consider is ethics. Ethics, or moral codes of conduct, while being culturally constructed, also serve as psychological frameworks that subliminally inform us that we should treat others the way in which we would want to be treated ourselves. How people understand this principle varies from culture to culture according to complex rules and standards of behavior, relations and interpretation. Yet, through the cultural frames of ethics, humans experience phenomena as either good (beneficial to them and/or to others) and evil (detrimental to them and/or to others). Through juxtapositions of good and evil, human emotions are wired to experience phenomena in either positive or negative ways, and to various extents depending on the semantic (meaning) value of what is being experienced.

Thus, ethics are an inseparable part of the emotional experience of technology design. Connections between ethics and emotions have been discussed in European philosophy for centuries. While the evaluation of the cultural production of technology is highly reliant on higher order cognitive-affective processes, the underlying and heavily instilled ethical principles that humans are raised by entail an attachment between worldly phenomena (technology), emotions and ethics that affects people at primal cognitive-emotional levels. For instance, encountering a rifle in one's living room may instinctively trigger fear depending on the circumstances. For this reason, understandings of ethics, how they are constructed and operate, and more importantly how they and technological interactions existing in holistic life ecosystems is highly valuable within technology design. From this point of view, considerations for technoethics provides a sense making extension to basic human technology interaction and design research.

Chapter 9
That Crazy World We'll Live in—Emotions and Anticipations of Radical Future Technology Design

Rebekah Rousi

Let's break out of the horrible shell of wisdom and throw ourselves like pride-ripened fruit into the wide, contorted mouth of the wind! Let's give ourselves utterly to the Unknown, not in desperation but only to replenish the deep wells of the Absurd!
(Marinetti 1909 [2016], 1)

Abstract Humans behave towards and experience technological design in conflicting and contradictory ways. On the one hand, the very mention of the word 'future' conjures expectations of the radically new and unexpected. On the other hand, previous research has shown that people have a threshold for the level of change and the unexpected that they can cope with. Their expectations are dominated by mental images of familiar associations with what has been previously associated with the future. As a rule, humans cope with incremental changes, yet have difficulty accepting the entirely unfamiliar. This makes it harder to imagine a future of radical technology design and interactions, particularly when attempting to predict possible emotional outcomes. This chapter describes the emotional balance between the familiar and the unfamiliar in design creations, which has also been observed in theories such as the Most Advanced Yet Acceptable (MAYA) theory. The MAYA theory emphasises the complexity and irony of incremental versus radical changes in renewing design language and technological systems for the future. By carefully observing classical cases of previous game-changing technological innovations, including their hype and acceptance curves, a model is proposed that illustrates how a radical design future may be achieved by tapping into emotional, ideological and interactive logic, rather than formalistic (material-based) design choices. This chapter highlights the role that culture and cultural discourse play in cognition and emotions when considering future technology design in terms of 'thinking outside the box'.

R. Rousi (✉)
Faculty of Information Technology, University of Jyväskylä and Gofore PLC, Jyväskylä, Finland
e-mail: rebekah.rousi@jyu.fi

© Springer Nature Switzerland AG 2020
R. Rousi et al. (eds.), *Emotions in Technology Design: From Experience to Ethics*, Human–Computer Interaction Series,
https://doi.org/10.1007/978-3-030-53483-7_9

9.1 Introduction—We Know the Robots Are Coming

The past has occurred, the present is happening and the future is waiting to unfold. The inevitable unknown may conjure images of systemic heaven or technological hell, depending on how the subject is framed. While the future is often thought of as unknown, radical, troubled or different, it is not known how it is anticipated and imagined. That is, knowingly or unknowingly, individuals hold a socially and culturally constructed stylised idea of what the future will look like (Inayatullah 1990)—and perhaps even feel, smell, sound and taste like. The future holds a design language of its own. Strong emotions are also associated with the design language of future representations. The emotional qualities of futuristic design and technologies are highly dependent on context and associations that are continually emphasised through popular culture and politics.

The notion of 'futuristic' design conjures images of the iconic spaces and structures imagined by the great architect Eero Saarinen, who designed New York City's JFK Airport, with its spacecraft-like wing-shaped roof, and the 192-m high stainless steel Gateway Arch in St Louis, Missouri, is the tallest in the world (National Park Service 2010). Certain aspects of his style have been adopted by modernist designers such as Ludwig Mies van der Rohe and Walter Gropius and the international style in general (Art Story Foundation 2019). Other sources of inspiration in Saarinen's work can be traced back to sixteenth century Vicenza, if not directly to the designs of the period, indirectly through other architects, artists and designers who were influenced by the forms of the era. Perhaps contrary to the chapter's opening quote extracted from the Futurist Manifesto (Marinetti 1909[2016]), Saarinen described the purpose of architecture as sheltering and enhancing "man's life on earth and to fulfil his belief in the nobility of his existence" (cited in Central Intelligence Agency 2017). French dramatist Eugène Ionesco (1957) described the 'absurd' as "devoid of purpose... Cut off from his religious, metaphysical, and transcendental roots, man is lost; all his actions become senseless, absurd, useless" (cited in Esslin 2009, 23). Saarinen's work serves a physical and existential purpose. His design language provided spatial effects that protected people from the effects of existing in an ever more complicated modern world (Art Story Foundation 2019).

Thus, based on the sensibilities of work by architects such as Saarinen, the aesthetic language of future design may be observed in smooth shiny curvaceous objects, gadgets, transportation and architecture. On this basis, design psychology theories such as the Most Advanced Yet Acceptable (MAYA, Hekkert et al. 2003; Baha et al. 2012) and scholarship on design typicality (Mayer and Landwehr 2018) argue that people are inclined to emotionally and aesthetically prefer prototypical designs (even of the unknown or supposed 'new') over atypical designs (Graf and Landwehr 2017; Hekkert et al. 2003; Veryzer and Hutchinson 1998). Thus, the technological future is typified by what people already understand to be the future. Minimalism plays a major role in 'future aesthetics', or the 'aesthetics of the new', with striking white spaces and transparency mitigated through an abundance of glass, steel and controlled colour palettes (Macarthur 2002).

On an immaterial level, the notion of 'future' appeals to a range of emotions, connected to both utopian and dystopian visions and typified by a compound of cognitive–emotional processes that derive from individuals existing in perceived uncertainty. There is almost never one without the other. For while Saarinen's creations form the basis of imagining a safe, clean and organised future, the rationale behind them is to protect people from the futuristic technological mess they have created. Thus, there is a sense of *human technology*—designing technology that protects and promulgates human values. Given the schizophrenic nature of the imagined technological future, this chapter seeks to achieve two main goals. First, it argues that the future of the so-called radical technological design is already known and has been developed incrementally over the centuries. Thus, people's notions and expectations of the future are already set in a pattern of futuristic aesthetics to which designers are expected to aspire for years to come.

Second, the chapter explores a technologically driven human side, characterised by madness or craziness (deviations from the so-called typical mental activity) that is coupled by emotions associated with the future. The Futurist cultural movement established during the early 1900s is described in relation to its radical and somewhat untamed discourse on the future. Futurism could be said to explicitly reject the human-centred ideologies of modernist architects such as Saarinen. Certainly, the mindset that Futurism promulgated could be described as a counter to the research claims made in relation to theories such as MAYA. Futurism was fuelled by rebellion and thus charged with aroused emotions that were then mobilised for cultural, social and political propagandist purposes. Futurism was not only about immaterial sentiments about the future; rather, it was interwoven with technology, human relationships to technology and a deliberate instrumentalisation of madness—once again representing a break with a temporal and/or nostalgic tradition of norms. Futurism was an attempt to move unguarded, yet not vulnerably, towards an unknown future. These insights are significant when reflected against recent trends in design research that have observed the dynamics between user-centred design and the design imagination (design intention).

As Mattelmäki et al. (2014, 73) point out, there is an 'empathy trap' in which designers are too concerned about the information users are explicitly offering, which more often than not resonates with the design typicality of various technology types and spatio-temporal contexts. This ignores the potential of what the designer's imagination has to offer regarding future design directions. Overcoming bottlenecks in design thinking is crucial when disrupting norms in design language, trends, processes and innovation. Thus, "[i]f designers are not vigilant, the attempt to be empathic might articulate popular reflections instead of innovating more radical futures" (Mattelmäki et al. 2014, 73).

9.1.1 Futurism

The Futurist cultural movement focused on capturing and expressing new ways of experiencing the world that rapid technological developments afforded (Dominiczak 2013). This has been described in terms of representing the 'new face' of the world as seen in relation to (and through) emerging technologies (Bell 2007; Gualdoni 2009). The significant cultural products produced in the Futurist movement were verbal manifestos. Thus, language and the symbolic poetry of Futurist pioneer Filippo Tommaso Marinetti were the technologies Futurists relied on to construct emotional images of the future. At the heart of Futurism was a fascination with technology, science, machines and movement (Dominiczak 2013).

The emotions portrayed and constructed in the Futurists' cultural rhetoric were driven by Marinetti's own emotional experience of Italian culture. Futurist rhetoric featured an aggressive emotional quality that promulgated the beauty of speed as well as a desire for human aggression through technological warfare (Dominiczak 2013; Marinetti 1909/2016). This is not surprising, given Marinetti's support for Mussolini's fascist regime. Thus, technology, emotions and culture were tightly interlinked and operationalised for political and warfare purposes. To instil the emotions associated with culture, cultural products and more traditional cultural technology museums, for instance, were deemed 'vicious' (Boccioni 2011; Boccioni et al. 1882-1962/1992).

The Futurist approach to science and technology was that of societal transformation. Moreover, technological objects were viewed in terms of their unity with the surrounding environment. Intensively masculine driven, Boccioni used the term 'interpenetration' in 1910 (cited in Harrison et al. 1993) to describe how science was providing the technological input to produce objects (and systems) to better serve people's material and intellectual needs. From the perspective of cognitive science, technology can be seen to be intrinsically connected with cognition and the development and expression of human thought—what Shaffer and Clinton (2006) have called 'toolforthoughts'. Technology can be considered as expression and embodiment of shifting and transforming conditions (social, political, economic, environmental, etc.) of human life, as expressed by Boccioni (cited in Harrison et al. 1993) and as witnessed in the evolution of technology-supported actions described in the theory and method of life-based design (LBD) (Leikas 2009; Saariluoma and Leikas 2010). Such actions and behaviours in people's everyday lives rely on technology. These change over time in relation to an individual's life circumstances as well as the ever-changing nature of society and culture.

9.2   Futurism, Madness and Emotions

When Futurism was conceived, Europe was in the midst of radical technological and societal changes (Pobuda 2017; Sconce 2011). Industrialisation was taking place at an overwhelming speed. Electric lights began to line city streets. Electrical and electromagnetic induction-powered machines began to permeate industry and eventually homes. X-rays made it possible to see inside the human body without invasive surgery, and the world became ever more connected via telecommunications and mass media—increasing long-distance communication without the need to physically travel. While these technological advancements seem wondrous, they came with social and psychological repercussions. Changes in the sounds of the urban environmental scape, in addition to all the other sensory stimulations that began to bombard city inhabitants, were said to contribute to the condition of neurasthenia—a weakness of the nerves that triggered anxiety, depression and fatigue (Gijswijt-Hofstra 2019; Sconce 2011; Taylor 2001).

In addition to the concrete effects that technological advancements were having on the brain and body, technological developments also spurred the imagination in terms of its ability to allow people to transcend from the natural to the supernatural. For instance, in the early 1900s, there were technological aspirations to communicate with the dead as well as wordlessly from one person to another (Chessa 2012; Pobuda 2017). These issues have recently gained momentum in public technological and scientific discussion, for example, Elon Musk's[1] Neuralink brain implant initiatives (Hitti 2019) and the availability of big data and internet profiles of the deceased (Berman 2018; Economou 2019) that enable people to seemingly talk to the dead through artificial intelligence (AI). Tanya Pobuda (2017) ponders whether the Futurists were fascinated with technology or indeed magic.[2]

The turn towards magic and the occult compounded the Futurist movement's deviation from the Catholic Church and the increased incidence of anxiety (neurasthenia) in society caused by the proliferation of high-speed technology, artificial lights and machine-induced soundscapes (Chessa 2012). Marshall McLuhan, who pioneered the term 'the medium is the message', postulated that the technological network that is increasingly connecting and taking over human society is an extension or live model of the human nervous system. This form of thinking uncannily predicts what is currently happening in terms of initiatives such as cognitive computing and the *Google cognitive–experiential takeover*. Thus, from an aesthetic perspective, scholars have connected Futurism to madness based on a combination of psychological (impact of technology on the mind and body), spiritual (rejection of religion and move towards the mystical pseudo-scientific connections of technology and the occult) and political (disruption of traditions, material and collective memory, and promulgation of political and economic agenda through technological speed and magic) principles. Once again, the modern-day informal Futurist Musk can be described as mad, not only simply in his drive to utilise information technology to achieve *magic* but also in his understanding of himself as a unique power unit in the society of the connected mind.[3] This connected mind—as it should be understood

---

[1] Musk could be described as a bold Futurist of our time, embracing many of the futuristic ideas of the past and present to forge a technological and business future.

[2] General Magic is a defunct technology company responsible for many of today's 'smart' innovations. It is briefly discussed later in the chapter.

[3] See, for example, https://www.wired.com/story/elon-musk-tesla-tweets-struggles/.

from recent developments such as cognitive computing—needs to be characterised not just in terms of a simplistic notion of cognition, but rather in relation to dimensions, layers and states of emotion, framing, intention and apperception (the process of creating mental representations of the world and its phenomena based on the intermingling of a variety of sensory and mentally based information, see, e.g. Saariluoma 2015).

Emotions play a key role in a number of cognitive functions ranging from attention and memory to sentiments and sense-making (experiential recall and logic, see Rousi 2013). However, emotions play a significant role in how people experience Futurism and expectations of the future as well as future design. Moreover, in terms of overall well-being, emotions constitute the greatest contributors and experiential qualities of mental illness (Melges and Fougerousse 1966). While there are numerous theories of emotions, particularly basic emotions, Carrolle E. Izard (2013) provides a useful categorisation of the fundamental emotions of anger, disgust, fear, shame, contempt distress, surprise, joy and interest. These emotional terms are relevant when considering human reactions to future design and especially the future in general. In his work on emotions in anxiety and depression, Izard (2013) delved deeper to describe emotions as existing in patterns, or in combination with, particular states and circumstances. The reason for considering emotions in combinations, or as complex, stems from the understanding that humans experience multiple emotions at a given point in time. In specific states such as anxiety and depression, patterns can be observed in combinations of emotions that pertain not only simply to the fundamental emotions but also to somatic (bodily) and cognitive components.

Thus, when observing madness in relation to Futurism and future design experience, both madness and emotions can be used to describe human responses to environmental influences (Theodorou 1993). Moreover, the strong (aroused) experience and expression of emotions such as anger, fear and even joy and desire have often been considered interchangeably with madness (De Sousa 1990). Likewise, emotions have often been said to threaten rational thought processes. Yet, due to the integral role that emotions play in thought and cognition, it can be said that they represent rationality (De Sousa 1990; Rousi 2013). Emotions provide the platform, basis or frame through which information is received (perceived) and processed. They also facilitate the connection of information chunks depending on the concerns (Frijda 1988, 1993)—objectives, goals and interests—of the thinker. Emotions in the context of embodied (somatic) cognition also regulate the type of cognitive processing from primal or lower order (immediate and often somatic) responses to higher order, secondary and associative processing (Brave and Nass 2007; Hekkert 2006).

Many emotions are considered to be associated with the so-called madness, manifesting in social, physical and cultural expressions. Thus, the idea of a radical future and radical future design may indeed be more of an emotional response to uncertainty as well as sociocultural hype, than to design and innovations themselves. In order for future design to *exist*, it must be recognised as such.

9.3 MAYA and the Stylistic Language of the Future

As with user experience design and the need to understand *how* to design in order to encourage and enhance specific emotional responses, a 'vision' for (and of) the future is incredibly important when attempting to approach and promote radical innovation (Chandy, Prabhu and Antia 2003). The stumbling block, however, is the issue of design typicality (Mayer and Landwehr 2018) and people's preference for familiarity (Hekkert et al. 2003). The story of the ill-fated General Magic company illustrates how if an idea or design is too new, or arrives too early, before the public's (market's) imagination, even the most potentially desirable product will fail (Kerruish, Maude and Stern 2018). General Magic, an electronics and software company founded in 1990, has been described as the "[m]ost important dead company in Silicon Valley" (Kanellos 2011). It developed and released devices with touchscreens, touchscreen controllers, multimedia emails, USBs, streaming televisions, e-commerce, personal digital assistants and networked games, which were beyond consumers' imaginations in the 1990s.[4] Since the public failed to appreciate what the technology afforded, or were sceptical of its capacity and reliability, sales floundered.

Thus, the timing of the technology was too early as people could not link the designs to their own lived experience: they did not see how the technologies could fit into their everyday ways of being.[5] For example, the company developed the Sony Magic Link, an early predecessor of the smartphone that offered messaging, word processing, spreadsheets, databooks, faxing, address book, calculator, calendar and accounting software and even served as a remote control. In on-the-spot interviews with people on the street around the time of its launch in 1994, people stated that they did not want such a device and could not think of using it (Kerruish et al. 2018). At the time of release, this mode of undertaking daily activities did not fit into people's life systems (routines and behavioural patterns), which involved physical branches and face-to-face transactions.

The main idea behind the smartphone, or Magic Link, dates back centuries to imagining items such as magic wands.[6] These all-in-one objects, which resemble remote controls, enable functions from transportation to transformation. Modern equivalents include the wearable gadgets of Maxwell Smart (Agent 86 from the television show *Get Smart*, 1965–1970) and James Bond (Agent 007, 1953 to present). Based on the experiences and familiarisation gained using these forms of imagined technology through popular culture, it could be assumed that in the 1990s wearable mobile technology in the form of shoe phones, dart gun neckties, audio recording cameras and clothing brush communicators, for example, in practice would have been mature ideas. Yet, when presented with the real-life opportunity to engage with and own a mobile computing device that could replace the need to physically

---

[4]These innovations were also arguably overshadowed by the deployment of the World Wide Web (Kerruish, Maude and Stern 2018).

[5]To understand this phenomenon, see Leikas (2009) or Saariluoma and Leikas (2010).

[6]Incidentally, 'magic wand' is the name given to numerous mobile information technology projects and innovations (see, e.g. Hansen et al. 1997; Ouchi et al. 2005).

travel to various locations, utilise multiple devices and deal with people, consumers were not ready. This raises the question of how ready today's global population is for technologies such as teleportation or even autonomous and learning domestic robots.

Thus time can be understood to play a role in individuals' willingness to engage in and adopt technological innovations. This is one aspect to consider from the perspective of the MAYA theory (Hekkert et al. 2003), particularly the maturation of familiarity with designs and their relevance through cognitive occurrences such as the mere exposure effect—a theory that posits that multiple exposures to phenomena increases the likelihood of its acceptance (Bornstein 1989; Zajonc 1968). Furthermore, *meaning* in relation to lived experience (Saariluoma and Leikas 2010), in terms of what designs afford a person to do and achieve within how they experience their life, is a component of MAYA. This is where the semantics (meaning) in terms of affordance (Gibson 1979)—what the technology allows users to perform—of technology derived from the semantic systems of past and present sociocultural regimes (Baha et al. 2012; Loewy 1951).

Timing may also influence when people believe certain technological capabilities should be possible to realise—that is, their temporal notions of technological experience and when they expect such technological capabilities to be possible. This involves living life in relation to technology with the expectations of what should be 'to hand' (Heidegger 1962)—and when—regarding their own life and techno-generational timeline (Canas-Bajo et al. 2016; Sackmann and Winkler 2013). People hold set notions, or mental representations of the past, present and future, throughout their lifetimes. This aspect qualifies the Futurist rhetoric of speed and their plights to leave the past—and even the present—behind (Marinetti 1909[2016]) in order to embrace the pace of technology. The underlying idea of the sociohistorical Futurist perspective was to break with mental models instilled by previous cultural traditions in Italy.

Returning to the style or design language of future technology, we refer to design typicality, which is recognised as future technological design. Thus, there are expectations of what radical future technologies will look and be like based on what is understood through design typicality—or 'the goodness of example' (Barsalou 1985; Hekkert et al. 2003; Loken and Ward 1990). Regarding future technology, design typicality is constantly being manifested and instilled through popular culture, popular science and other media discourse 'predicting' the future. In other words, popular culture helps familiarise the public with the language and innovations of the future. For instance, in a blog article published in 2015, Dean Evans stated that over the next 20 years, there will be a huge leap in the technologies and structures of the world. While some things will change dramatically beyond recognition, others will remain the same. A vision of the future that has been developing over the past 100 years or so that is very pronounced in today's societal, economic, political and design discourse, is one in which everything—from nutrition and health (see, the biohacker movement for instance, as well as basic forms of food production), to work and relationships—is mitigated by overt forms, expressions, language, logic and aesthetics associated with science (e.g., biotech), information technology and overall connectivity.

Another issue that affects the direction of future information technology design is the technology's proximity to the body (Wong, featured in Floyd 2018). While mainstream consumers are adapting to the idea of wearable computing, information technology is under development that will be implantable and ingestible. Thus, the outward appearance may play an instrumental role in convincing an individual to engage in the technology today, yet its deployment and use will have nothing to do with the design aesthetics of form in the future. It will be more about the feel, function and augmentation enabled by these technologies, and how the individual feels as an entity, or a citizen, as to whether or not people are satisfied with and positively experiencing these technologies. In other words, the future user experience will be about how technology affects the user's physical, embodied emotional experience. A recent study by Rousi and Renko (2020) explores participants' opinions on cognitive enhancement technology. The technological designs were divided into three groups—environmental, on-body (exo) and in-body (endo); the results reflected an increase in negative emotional reactions towards endo technology. Thus, current research suggests that today's information technology consumers may not be ready for the types of design we cannot see and feel, which invades the human body and has the potential to access and control more of individuals than can even be considered. For instance, Elon Musk has proposed the use of brain implants to eliminate the need for verbal language.

9.4 Embodied Emotions, Future Technology and the Radical Tech Conclusion

The Futurist movement was cultural, discursive and intrinsically connected to the arts. Art, in turn, is one of the oldest known forms of technology. The word 'technology' in fact comes from the Greek term 'techne' that describes art and craft, in combination with 'logos'—speech and word, or discourse (Buchanan 2019). From this perspective, the design form itself is the discourse, a discussion of the present and projections of the future, through connections to the past. Futurism was instrumental in the Italian dictator Mussolini's fascist political propaganda; it promoted speed and the desire, or need, to let go of traditions and the past as well as technological warfare. When considering the close connection that Futurism (a cultural movement) had with science, technology, politics (fascism) and violence, it is little wonder that strong basic emotions such as fear, anger and disgust (see, e.g. Ekman 1992 and Prinz 2004 for arguments on basic emotion) can be induced by thoughts of future technology. The technology was literally embodied in weaponry and warfare; several Futurists (including Boccioni) died in World War I (Dominiczak 2013). Thus, trust in and emotions towards future technologies (or the lack thereof) could very well be rationally founded (Saariluoma et al. 2019).

Thus, the public may have a right to be emotionally drawn to the familiar soft curves and shelters, similar to Saarinen's minimalist futuristic designs, because throwing all tradition to the wind, as Marinetti (1909[2016]) chants, can lead to technological violence and warfare. Expressions of future technology, combined with a history of technological violence promulgated by popular culture, incite rational emotions and states of fear and anxiety about future designs that are profoundly embodied by nature (De Sousa 1990; Saariluoma et al. 2019). According to the principles of MAYA (Hekkert et al. 2003; Baha et al. 2012), people do not accept the unknown and entirely unfamiliar, as they cannot assimilate them into their previously learned information (apperception, see, e.g. Saariluoma and Rousi 2015). Yet, what is known and understood can often trigger negative emotions posed by the potential danger of emerging technologies as understood through past cultural, social and physical experiences.

Contemporary society is plagued by concerns about mental health just as in the time of the Futurist movement, particularly in relation to technologically induced mental and social disorders. Neurasthenia—the experience of nerve weakness, anxiety, depression and fatigue (Gijswijt-Hofstra 2019; Sconce 2011; Taylor 2001)—is as prevalent today as it was in 1902 when Segalen employed the term to describe people's adverse cognitive, emotional and sensory reactions to the rapid technological developments happening in their lived environment (Goulet 2013). There is currently a considerable discussion about the 4G Zombies and the 5G 'Madzone' (Edwards 2019). It is difficult to tell whether the fear, apprehension and anxiety associated with mental images of radical future designs are culturally induced by past imaginings and circumstances, or whether they are technologically induced by current and emerging technological use. With contemporary, Futurists such as Elon Musk pressing forward with embedding information technology into the human body, and especially the nervous system, one may ponder whether extreme emotional reactions to what is already *known* about these types of future technologies should signal to designers to take the reins of development and steer them towards a sheltered and possibly brighter future. For it seems, particularly when the materialism of the immaterial and the intelligence of the artificial are concerned, the lunatics really are running the asylum (Cooper 2004).

References

Art Story, The (2019) Eero Saarinen—Finnish-American architect and designer. Movements and styles. https://www.theartstory.org/artist-saarinen-eero.htm. Accessed 13 Feb 3029

Baha E, Lu Y, Brombacher A, van Mensvoort K (2012) Most advanced yet acceptable, but don't forget. In: DS 71: Proceedings of NordDesign 2012, the 9th NordDesign Conference, Aarlborg University, Denmark

Barsalou LW (1985) Ideals, central tendency, and frequency of instantiation as determinants of graded structure in categories. J Exp Psychol Learning, Memory Cognition 11(4):629–654

Bell J (2007) Mirror of the world. A new history of art. London, Thames and Hudson, pp 374–390

Berman R (2018) The chance to text with the dead via AI is creepy or wonderful. *Big Think*. https://bigthink.com/robby-berman/the-chance-to-text-with-the-dead-via-ai-is-creepy-or-wonderful. Accessed 21 July 2019

Boccioni U (2011) Manifesto of Futurist painters. In: Danchev A (Ed) 100 artists' manifestos. From the futurists to the stuckists. London: Penguin Books, pp 14–27

Boccioni U, Carrà C, Russolo L, Balla G, Serverini G (1882–1962/1992) Futurist painting: Technical manifesto. In: Harrison C, Wood P (eds) Art in theory 1900–1990 (149–152). Oxford, UK, Blackwell

Bornstein RF (1989) Exposure and affect: overview and meta-analysis of research, 1968–1987. Psychol Bull 106:265–289

Brave S, Nass C (2007) Emotion in human-computer interaction. Handbook of human–computer interaction: fundamentals, evolving technologies and emerging applications. Hillsdale, NJ: Lawrence Erlbaum, pp 77–92

Buchanan RA (2019) History of technology. *Encyclopaedia Britanica*. https://www.britannica.com/technology/history-of-technology/Technology-in-the-ancient-world. Accessed 31 Oct 2019

Canas-Bajo J, Leikas J, Jokinen J, Cañas JJ, Saariluoma P (2016) How older and younger people see technology in the Northern and Southern Europe: Closing a generational gap. Gerontechnology 14(2):110–117

Chandy RK, Prabhu JC, Antia KD (2003) What will the future bring? Dominance, technology expectations, and radical innovation. J Marketing 67(3):1–18

Chessa L (2012) Luigi Russolo, futurist: Noise, visual arts, and the occult. University of California Press, Berkeley

Central Intelligence Agency (2017) Eero Saarinen—a place in architectural history. https://www.cia.gov/news-information/featured-story-archive/2017-featured-story-archive/eero-saarinen-a-place-in-architectural-history.html. Accessed 19 Mar 2019

Cooper A (2004) The inmates are running the asylum: Why high-tech products drive us crazy and how to restore the sanity, vol 2. Sams, Indianapolis, IN

De Sousa R (1990) The rationality of emotion. MIT Press, Cambridge, MA

Dominiczak MH (2013) Technology and emotions: the futurists. Clin Chem 59(2):453–455

Economou E (2019) Dead but not gone—how Facebook AI is bringing deceased friends back into your life again. *Towards Data Science*. https://towardsdatascience.com/dead-but-not-gone-how-facebook-ai-bringing-deceased-friends-back-3ea8d865d01b. Accessed 21 July 2019

Edwards C (2019) The 5G Electromagnetic 'Mad Zone' Poised to Self-Destruct: The 5G 'Dementors' Meet the 4G 'Zombie Apocalypse'. *Global Research*. https://www.globalresearch.ca/5g-electromagnetic-mad-zone-poised-self-destruct-5g-dementors-meet-4g-zombie-apocalypse/5689176. Accessed 15 Sept 2019

Ekman P (1992) Are there basic emotions? Psychol Rev 99(3):550–553

Esslin M (2009) The theatre of the absurd. Vintage, New York

Evans D (2015) What will life be like in 2035? *Grey is the new black*. http://www.suncorp.com.au/super/grey/your-future/what-will-life-be-2035. Accessed 18 Aug 2019

Floyd C (2018) Brian Wong. In: 50 years we'll have 'robot angels' and will be able to merge our brains with AI, according to technology experts [Video]. https://nordic.businessinsider.com/we-asked-6-experts-what-the-world-will-look-like-in-50-years-time-mwc-2018-3. Accessed 18 Aug 2019

Frijda N (1988) The laws of emotion. Am Psychol 43(5):349–358

Frijda N (1993) Moods, emotion episodes, and emotions. In: Lewis M, Harviland JM (eds) Handbook of emotions. Guilford Press, New York, pp 381–404

Gibson JJ (1979) The theory of affordances. *The ecological approach to visual perception*. Houghton Mifflin, Boston, MA

Gijswijt-Hofstra M (2019) Neurasthenia. In: Albrecht G (ed), Encyclopaedia of Disability. http://sk.sagepub.com/reference/disability/n575.xml. Accessed 19 Mar 2019

Goulet A (2013) Neurosyphilitics and madmen: the French *Fin-de-siècle* fictions of Huysmans, Lermina, and Maupassant. Prog Brain Res 206:73–91

Graf LK, Landwehr JR (2017) Aesthetic pleasure versus aesthetic interest: The two routes to aesthetic liking. Frontiers Psychol 8(15)

Gualdoni F (2009) Futurism. Skira Editions, Milan

Hansen H, Hadjiefthymiades S, Immonen I, Kaloxylos A (1997) Description of the handover algorithm for the Wireless ATM Network Demonstrator (WAND). University of Athens, Athens

Harrison C, Frascina F, Perry G (1993) Primitivism, cubism, abstraction: the early twentieth century. Yale University Press, New Haven

Heidegger M (1962) Being and time. (Trans: Macquarrie J and Robinson E). New York, Harper & Row

Hekkert P (2006) Design aesthetics: principles of pleasure in design. Psychol Sci 48(2):157–172

Hekkert P, Snelders D, Van Wieringen PC (2003) 'Most advanced, yet acceptable': typicality and novelty as joint predictors of aesthetic preference in industrial design. Br J Psychol 94(1):111–124

Hitti N (2019) Elon Musk's Neuralink implant will 'merge' humans with AI. De Zeen. https://www.dezeen.com/2019/07/22/elon-musk-neuralink-implant-ai-technology/. Accessed 21 Aug 2019

Inayatullah S (1990) Deconstructing and reconstructing the future: Predictive, cultural and critical epistemologies. Futures 22(2):115–141

Ionesco E (1957) Dans les armes de la ville. Cahiers de la compagnie Madeleine Renaud-Jean-Louis Barrault. Gallimard, Paris

Izard CE (2013) Patterns of emotions: a new analysis of anxiety and depression. Academic Press, Cambridge, MA

Kanellos M (2011) General magic: the most important dead company in Silicon Valley. Forbes. https://www.forbes.com/sites/michaelkanellos/2011/09/18/general-magic-the-most-important-dead-company-in-silicon-valley/. Accessed 13 Mar 2019

Kerruish S, Maude M, Stern M (2018) General Magic the movie [Documentary Film]. Spellbound Productions

Leikas J (2009) Life-based design: a holistic approach to designing human-technology interaction. VTT, Helsinki

Loewy R (1951) Never leave well enough alone. Simon and Schuster, New York

Loken B, Ward J (1990) Alternative approaches to understanding the determinants of typicality. J Consumer Res 17(2):111–126

Macarthur J (2002) The look of the object: Minimalism in art and architecture, then and now. Architect Theory Rev 7(1):137–148

Marinetti FT (1909/2016) The founding and manifesto of futurism. Art Press Books, p 23

Mattelmäki T, Vaajakallio K, Koskinen I (2014) What happened to empathic design? Des Issues 30(1):67–77

Mayer S, Landwehr JR (2018) Objective measures of design typicality. Des Stud 54:146–161

Melges FT, Fougerousse CE Jr (1966) Time sense, emotions, and acute mental illness. J Psychiatr Res 4(2):127–139

National Park Service (2010) National historic landmarks program: gateway arch. National Historic Landmarks Program. https://www.webcitation.org/5uynJZKkF?url, http://tps.cr.nps.gov/nhl/detail.cfm?ResourceId=2017&ResourceType=Structure. Accessed 14 Dec 2018

Ouchi K, Esaka N, Tamura Y, Hirahara M, Doi M (2005). Magic Wand: an intuitive gesture remote control for home appliances. In: Proceedings of the 2005 international conference on active media technology

Pobuda T (2017) Of madness and metaphysics: Italian Futurism at the dawn of the 20th century. https://www.academia.edu/35717394/Of_Madness_and_Metaphysics_Italian_Futurism_at_the_Dawn_of_the_20th_Century?auto=download. Accessed 9 Mar 2019

Prinz J (2004) Which emotions are basic. In: Evans D, Cruse P (eds) Emotion, evolution, and rationality. Oxford University Press, Oxford, pp 69–88

Rousi R (2013) From cute to content—user experience from a cognitive semiotic perspective [Doctoral dissertation]. Jyväskylä Studies in Computing 175. Jyväskylä, Finland, University of Jyväskylä

Rousi R, Renko R (2020) Emotions toward cognitive enhancement technologies and the body—attitudes and willingness to use. Int J Human Comput Stud 143: 102472

Saariluoma P (2015) Foundational analysis: Presuppositions in experimental psychology. Routledge, London

Saariluoma P, Karvonen H, Rousi R (2019) Techno-trust and rational trust in technology—a conceptual investigation. In: Barricelli B et al. (eds) Human work interaction design. Designing Engaging Automation. HWID 2018. IFIP Advances in Information and Communication Technology, 544. Springer, Cham

Saariluoma P, Leikas J (2010) Life-based design-an approach to design for life. Global J Manage Business Res 10(5):27–33

Saariluoma P, Rousi R (2015) Symbolic interactions: towards a cognitive scientific theory of meaning in human technology interaction. J Adv Humanities 3(3):310–324

Sackmann R, Winkler O (2013) Technology generations revisited: the internet generation. Gerontechnology 11(4):493–503

Sconce J (2011) Origins of the influencing machine. In: Huhtamo E, Parikka J (eds) Media archaeology: approaches, applications, and implications. University of California Press, Berkeley

Shaffer D, Clinton K (2006) Toolforthoughts: Re-examining thinking in the digital age. Mind, Culture Activity 13(4):283–300

Taylor RE (2001) Death of neurasthenia and its psychological reincarnation. A study of neurasthenia at the National Hospital for the Relief and Cure of the Paralysed and Epileptic, Queen Square, London, 1870–1932. British J Psychiatry 179(6):550–557

Theodorou Z (1993) Subject to emotion: exploring madness in Orestes. Classical Q 43(1):32–46

Veryzer RW Jr, Hutchinson JW (1998) The influence of unity and prototypicality on aesthetic responses to new product designs. J Consumer Res 24(4):374–394

Zajonc RB (1968) Attitudinal effects of mere exposure. J Pers Soc Psychol 9:1–27

Chapter 10
Aesthetic Well-Being and Ethical Design of Technology

Jaana Leikas

Abstract Aesthetics is a central quality attribute of a product. Research into the relationship between aesthetics in human–technology interaction and the well-being of older people is still in its infancy. In care homes, aesthetics can play a major role in creating a 'feeling of home', which is important when the transition to assisted living may involve multiple changes and losses that affect an older person's well-being. This chapter discusses the potential of aesthetic design to address older people's emotional well-being. Aesthetics in technology and technological environments provides a new ethical way of looking at valuable problems in design—meaningfulness in terms of personal and individual symbolic values and the harmonizing potential of artefacts to create a 'feeling of home'. Promoting the aesthetic well-being of older people in care homes (and in general) deserves further attention.

10.1 Introduction

One of the biggest challenges of modern societies is how to respond to the needs, preferences and lifestyles of an ageing population. Expectations of technology are high: it is seen as a means to enhance the quality of older people's everyday lives whether they live at home or in an assisted living facility. Several studies have recently examined older people's capabilities as technology users, including the difficulties they face when trying to adopt new technologies. Yet improving older people's *experience* of using technologies has been largely overlooked. Senior citizens often find modern technology to be complex, obscure, confusing and not aesthetically pleasing. At the extreme, it is stigmatizing and does not meet their emotional needs.

'Home' is a value-laden concept (Hayward 1975; Moloney 1997; Sixsmith 1986; Madden 1999) that is usually taken for granted (Buttimer 1980) even though it is often the most central place in one's life (O'Bryant 1983; Rioux 2005; Swenson 1998; Williams 2002, 2004). Home involves a place or space, but for many people,

J. Leikas (✉)
VTT, Tampere, Finland
e-mail: jaana.leikas@vtt.fi

© Springer Nature Switzerland AG 2020
R. Rousi et al. (eds.), *Emotions in Technology Design: From Experience to Ethics*, Human–Computer Interaction Series,
https://doi.org/10.1007/978-3-030-53483-7_10

it also entails feelings, practices and/or an active state of being in the world (Mallet 2004). For most people, home is a safe haven that is experienced not only by people in it but also from meanings and memories associated with the look and form of its artefacts. Carefully acquired furniture, paintings and textiles make a house or apartment 'home'. Over time, these artefacts can be thought of as a part of one's identity.

Many elderly people have lived in the same home for decades. Furniture passed down through generations, crafts created by loved ones and wedding and anniversary gifts trace an individual's life history and values and create a sense of rootedness and belonging. Their home is thus an integral and intimate part of their being. The passage of time and the memories embedded in the home tend to deepen the older adult's sense of rootedness (Dahlin-Ivanoff et al. 2007; Saunders 1989); when they lose their home, they also lose the place closest to their heart (Gillsjö et al. 2011). It is difficult for them to decide what to bring with them in this final move to a care home.

Aesthetic aspects of objects in the home can represent older people's personalities and lives and help them recall the past. A product's appearance has worth not only because it sends social and symbolic signals about the person but also because it creates happiness—well-being and harmony, belonging and identity. This is extremely important when an individual's memory starts to fail: artefacts can help encourage a harmonious and cosy feeling.

When various impairments force an elderly person to move to a barrier-free and sheltered (and often quite small) flat, decisions about the 'placement' and even its furnishings are often made by someone else—in the worst case, by a total stranger. Without the comfort of the familiar objects at home, an elderly person easily feels lost, as illustrated by the fictional case study below:

> Anna has lived in the same neighbourhood all her life, and in her current house for more than 40 years. Her birthplace is only 1 km away. These rooms and furniture carry the marks of her life. The chest of drawers and chairs tell stories about what has happened along her life's journey. They represent her family and inheritance and identity. Many generations have had the pleasure of using the sofa and armchair set, leaving their ornamented armrests shiny from years of use. And that kitchen cabinet Anna varnished with her husband. But now the time has come for Anna to move out of this home. Moving away is difficult. It involves being torn away from a bigger house to move into a smaller place, from the familiar to the unknown, from one's own peace and quiet into a care home or sheltered housing and life under the eyes of others. It requires letting go of one's own will and dear belongings, under the authority of others. Yet, living at home alone is taxing both physically and mentally, because loneliness is overpowering. The visits of the home care nurse for 'securing maximum support'—even if they take place three times a day—have not been enough to ease the need to chat with someone, to have a coffee together and recount memories. Sharing things around one's own coffee table would lift up one's spirits.

Familiar artefacts help individuals remember their past and hold onto their identity (Baddeley 1990; Schacter et al. 1993; Smith and Vela 2001). *Context-dependent memory* indicates that the more familiar a context, the better the human memory functions. This is important to keep in mind when a person is moving to an elderly home or service home. The old person is more capable of remembering things that are related to their life and activities when the reminiscing takes place in a familiar environment, and they are surrounded by familiar artefacts.

Part of the emotional value of, for instance, a piece of furniture, is the memories that are associated with it. Sentimental value may constitute a stronger argument for holding onto an artefact than its aesthetic value. In his 1990 novel *Fields of Glory*, Jean Rouaud writes about his grandmother moving from a large house in the south of France to a small flat. He illustrates how sentimental value is more important to his grandmother than the beauty of artefacts:

> The move from thirteen rooms to two meant parting not only with the accumulation of a lifetime but also with the bequests of earlier generations. More than asceticism, it was a sweeping away of memory. Still, it was grandmother's recollection of this past that drove her to keep two or three heirlooms, in particular a cumbersome, poorly designed work table, when she could have kept the attractive mahogany bookcase with the oval glass panes in the same space and to better advantage. But this work table was her mother, her grandmother, herself and every industrious woman in the family—it was a stele.

10.1.1 Values and Subjective Well-Being

Values have an aesthetic dimension—in addition to social, philosophical and practical ones—that help explicate individual conceptions of beauty. What is valuable to some people, even in terms of the acceptance of technology, is not necessarily important or valuable to all people. Individual values reflect societal demands and psychological needs. Values are learned and determined by culture as well as personal experiences and life situations. The life of an individual itself helps determine what is considered 'worth' in artefacts.

Values determine attitudes and the preferences of activities and are tightly linked to other concepts such as life satisfaction, morale, subjective well-being (SWB) and happiness (Amann et al. 2006; Diener and Eunkook 2000) Studies of SWB have explored what constitutes a good life and made it scientifically interesting. SWB represents the degree to which people are living their lives according to the values they hold dear and they want to follow in their life (Diener and Eunkook 2000). Subjective quality of life is an individual's assessment of the general positive or negative qualities of their life experiences. Many attempts have been made to label this phenomenon, including personal expressiveness and optimal experience, but the most easily understood term is 'happiness'.

Aristotle believed that happiness is the only goal that is an end in itself (Aristotle ND[1984], 1097b20). Happiness can be based on two kinds of dichotomies: 'life as a whole' and 'life aspects'. Overall happiness is the degree to which an individual favourably assesses the overall quality of her own life as a whole (Veenhoven 2012). Enjoyment of certain aspects of life—such as aesthetics—will contribute to overall life satisfaction (bottom-up effect) (Veenhoven 2012). Senior homes and care homes can enhance the happiness and subjective well-being of their inhabitants. Yet how many of them reflect older people's values and conceptions of beauty? How many inspire the elderly aesthetically?

10.1.2 Aesthetics in Relation to Technology

Can technology play a role in nurturing people's aesthetic well-being? In the field of human–technology interaction (HTI) design, Redström (2001) discusses three forms of use (originally introduced by Paulsson and Paulsson (1957)) that designers must acknowledge—practical, social and aesthetic. *Practical use* is involved, for example, when we use a saw to cut timber or a boat to cross a lake. Depending on the model, the boat can also have a strong symbolic value in social use, especially a motor or sailing boat. *Social use* concerns the symbolic values that different artefacts have in social contexts, in other words, the roles that things play in our social life. An example of this might be wearing a tie on different occasions. For some people, it is a symbol of respect for the occasion, whereas, for others, it might be a negative sign of a social class. *Aesthetic use* concerns reflective use such as choosing a product because of its beauty. In a way, aesthetic use goes beyond practical and social use, as it concerns our immediate perceptions of things in terms of likes and dislikes of products that we choose.

Emotions are an essential part of aesthetic well-being, as they create satisfaction and awareness of the artefact (Desmet 2003; Desmet and Hekkert 2002; Hassenzahl 2001; Holman 1986; Montague 1999; O'Connor 1997; Schütte 2005). The positive or negative appearance of aesthetic artefacts can have a significant effect on our consciousness, as it generates different emotions and sensations. Objects that are beautiful and harmonious create inner satisfaction and contribute to well-being in everyday life.

Emotions are closely related to feelings, though the term 'feeling' is often used in colloquial language to describe a broader concept than emotions (Oatley and Jenkings 1996). The psychological literature historically used 'affect', which is closely associated with emotion (Tomkins 1984). Emotions are also related to needs and motives—the quality attributes that a person tunes into through emotions (Norman 2004; Oatley and Jenkins 1996; Solomon 1993). Emotions influence what value we place on objects as well as our decisions about them. For instance, consumers are being shown to be more interested in aesthetic pleasure than usability (Bertelsen et al. 2004; Hassenzahl 2004; Tractinsky et al. 2000); artefacts that are considered beautiful are experienced as easier to use compared with equivalent 'uglier' versions (Monk and Lelos 2007). Likewise, products with low usability are experienced as less beautiful than those with high levels of perceived usability (Tractinsky et al. 2000). Beauty has been associated with the willingness to own or purchase an object (Saariluoma et al. 2013). Jordan (2000) talks about *pleasure* as an important factor in HTI design (see also Green and Jordan 2002). Pleasure can be defined as the emotional, hedonic and practical benefits associated with a product. Jordan argues that instead of stressing usability, products should be designed to be a joy to own and use. Pleasurability is thus not simply a property of a product but of the *interaction* between a product and a person. This relates to the notion of aesthetics of interaction, which is gaining attention in product design (Colombo et al. 2015).

Understanding how people mentally represent their interactions with objects at home illustrates how beauty is experienced, which in turn helps design technology that supports and improves aesthetic well-being. *Mental representations* have mental content that can be in the form of sensations, images, memory, thoughts, propositions, beliefs and emotions (Saariluoma 2003). Representational content represents something outside the object itself (Saariluoma 2003). In the conceptual framework of mental content, experience can generally be seen as the conscious aspect of mental representations (Rousi et al. 2010). Thus, designing for aesthetic well-being should focus on experience as the feelings, thoughts and emotions that manifest during interactions with a product (McKay et al. 2006)—that is, on the holistic experience and how people feel (and what they would like to feel) when possessing and using technology (Hassenzahl and Tractinsky 2006; Jordan 1998). This experience is created through a sense-making process that includes anticipating, connecting, interpreting, reflecting, appropriating and recounting (Wright et al. 2004).

10.2 How to Design Aesthetic Well-Being for Older People?

Designing technology that enhances (or at least does not contradict) the aesthetic values of older people requires understanding their everyday lives and what they consider beautiful in terms of people's values and 'worths'. This involves considering not only just the physical aspects of artefacts but also how older people mentally experience them. Components involved in the mental side of aesthetic experience include cultural, social, psychological and linguistic. When developing new technical solutions for a particular form of life, it is important that the technology fits and improves the value climate of the form of life (Leikas 2009). A home should enable older people to maintain not only their identity, integrity and sense of belonging but also their way of living and *beauty welfare*. It should ensure that the aesthetic features of their homes are in harmony with their conceptualizations of beauty.

The theory of salutogenesis—which involves focusing on health, rather than the disease itself—provides evidence that space contributes to health and well-being (Chrysikou et al. 2018). An interesting successful example of designing for older people is the memory village of Hogeweyk, a Dutch design-oriented facility for people with dementia. In Hogeweyk, houses are differentiated by lifestyle. Residents can pick from houses decorated with a distinct and very residential feel designed to replicate a 'genre' of lifestyle and create a link to the life they enjoyed before. These genres have different types of interior design, music, food and table settings (Medaesthetics 2014).

Additional efforts need to be made to understand older adults' experience of beauty at home and to efficiently incorporate this data into designs in order to maximize their well-being (Leikas 2009; Saariluoma et al. 2016). The demand for beauty should concern all kinds of technologies, including low or standard technologies as well as high technology—the innovative emerging technologies based on ubiquitous technology and artificial intelligence. This is challenging in the era of smart homes, where digitalization is taking place. When installing safety technology such as surveillance and monitoring devices in older people's homes, the home should remain 'a home' without the elderly person's flat turning into a 'virtual hospital' or 'monitoring command centre'. For example, a lacy doily placed on a computer in an elderly person's home may represent an attempt to reduce the conflict between her own concept of beauty and the external appearance of the modern device. This may be surprisingly important for how acceptable she considers technology.

As discussed above, for many people, the spirit of the home comes from the subjectively experienced beauty of the belongings inside. The designer should, therefore, consider how a new technology could maintain the memory-filled spirit of the old home and thus preserve or improve the aesthetic pleasure and well-being of its user. Co-design and narratives that illustrate the meanings and values embedded in objects help illustrate what kinds of artefacts older people want. Older people should be consulted directly. They should not just be invited to focus groups; their feedback should be carefully considered and incorporated into the design process.

10.2.1 Technology Generations

An interesting phenomenon in terms of beauty in technology is the rise of the analogue culture from the fringes of the digital mainstream. In some designs, analogue technologies have been 'rediscovered' in a new way such as installing new technology in old-fashioned covers. This approach could help older people enjoy new technology by exploiting context-dependent memory. For example, the Hulger Phone, designed by Nicolas Roope in 2005 (Fig. 10.1), can be adapted to a smartphone or computer. The name of the phone was inspired by the designer's grandfather, a lawyer who

Fig. 10.1 Hulger phone
(www.mocoloco.com/arc
hives/002896.php)

lived contentedly with the same phone, the same 1950 Opel car, and the same leather armchair for decades. He was of a generation and culture that resisted the wasteful churn of built objects becoming obsolete. He bought lasting products that improved with age. Roope wanted to create technological products with this integrity: not only to reduce the destructive demand on the world's resources but also to settle our souls, with products that genuinely make us feel contented and balanced.

The user interaction model of the Hulger phone relies on context-dependent memory and supports the idea of *technology generations*. The notion has become valuable in research into the effects of age in information and communication technology (ICT) use. Technology generations reflect the historical timing of computing innovations and their diffusion into the productive and cultural spheres, linked with the time period in which a cohort comes of age (McMullin et al. 2007). A technology generation is defined by the technology its members used when they were aged between 30 and 40. The understanding of how to use technologies (present and future) is built on the kind of knowledge that is typical of that cohort (Docampo Rama 2001). Accordingly, it could be assumed that the conceptions of beauty in technology can be linked to cohorts in the same way. Elderly people today belong to the technology generation that used rotary dial phones and typewriters. Do they consider these kinds of artefacts to have more aesthetic value than current ICT products?

10.2.2   Naturalness is Considered Beautiful

For many people, nature—as a part of everyday life—is a source of aesthetic experiences. Designers have attempted to copy the beauty of nature in their designs. Aesthetics and naturalness have often been considered related values, and homes have been decorated by respecting the naturalness and primal force of wood. The eminent Finnish architect Alvar Aalto (1898–1976) called nature a symbol of freedom and used the metaphor of nature's harmonious adaptability in his interior designs (Schildt 1997). As for so many Nordic people, nature was the essence of his personality and work. Aalto said that furnishings made from timber allow direct contact with the surrounding nature. Indeed, natural environments have been shown to promote good health (e.g. Maller et al. 2006) by reducing stress, speeding up the healing process and providing a distraction from discomfort. Environmental psychology has shown that aesthetic experience and environments are intimately linked (Steg et al. 2012). People enjoy better health and quality of life through contact with nature in their everyday living environment. This desire to enjoy the aesthetics and healing elements of nature does not disappear with age. Further research is needed to understand what nature-based interventions can and cannot do and how different mechanisms of 'nature as medicine' may be combined with our living environment.

The Finnish Lumo Video Security Phone (Fig. 10.2), developed for elderly people, is an example of a design that respects nature and adheres to the traditional Scandinavian values of interior design. Scandinavian design, found in the homes of many elderly people in the Nordic countries, has its origins in the characteristics of

Fig. 10.2 Lumo video
carephone (https://www.oul
umo.com/english.)

Nordic nature and includes organic forms, materials and natural patterns. The Lumo
Phone, which combines personal security, welfare monitoring and social interaction,
is designed in natural wooden covers that are appealingly simple.

Conceptions of beauty and the mental representations of design objects can be
explored by exposing latent user experience factors by applying the concept of
semantic differential to the design process. The study by Osgood et al. (1975) serves
as a classical example of research on how people use a semantic differential scale
consisting of affective word pairs, such as 'good–bad', to respond to different stimuli
when products are introduced to them. A similar methodology was applied in Kansei
engineering, in which the researcher aims to understand a product's semantic space,
i.e. the relationships between its different expressions (Nagamashi 1995). The word
'Kansei' refers to simultaneous feelings and images, i.e. the emotional and cognitive
levels of an experience.

Many older people would reject a product if it looks ugly or signals that it is
meant for the elderly. Traditional assistive technology is often quite stigmatizing in
this respect. For instance, 'senior apartments', 'senior phones' and 'senior packages'
suggest that they are meant for older—and somehow disabled—people (e.g., Ziefle
and Schaar 2017).

10.3   Discussion

This chapter is based on the hypothesis that the aesthetics of a care home or senior
community can enhance the subjective well-being of older people. Aesthetics should,
therefore, be considered when designing for the everyday lives of older people in

care homes. Further research is needed on how older adults can feel more 'at home' in care homes and the implications for aesthetic well-being. In order to understand which aspects to stress in the design, it is crucial to understand how older people mentally represent home and their favourite artefacts at home and what kinds of values they have followed in their life. Designers need to learn what kinds of attributes are associated with older people's conceptions of beauty at home and what meanings these attributes have. This information can be acquired with the help of various co-design methods and, for instance, content-based analysis of narratives.

Research is also needed on how the elderly value aesthetics in relation to other design attributes, such as usability, and how these would cohere with the design practices of new technologies. This points to the need to rethink technology design processes and to include older people as co-designers.

Older people's aesthetic well-being is interwoven with the idea of a good life. In the 1950s, beauty was considered one aspect of a good life, together with truth, character and fellowship (Fung and Lehmberg 2016). The concept of quality of life (QoL) reflects the philosophy of what is considered a good life (Nussbaum and Sen 1993). QoL refers to an individual's perception of her position in life, in the context of the culture and value system where she lives, and in relation to her goals, expectations, standards and concerns (WHO 1994). From a subjective perspective, QoL can be sorted between the outer and inner qualities—external and internal features—of life (Veenhoven 2012) and thus be regarded as a person's reaction with respect to these resources, according to her own values, goals and expectations.

The question of what makes a person's life better in terms of QoL arises in the course of a moral argument about our duties and obligations to make people's lives better or at least prevent them from being made worse (Scanlon 1993). Thus, in an ethical sense, technology designers have a duty to make the lives of older people better, for example, by creating an aesthetically emotional impact.

References

Amann A, Reiterer B, Risser R (2006) Life quality of senior citizens in relation to mobility conditions. Final Report. University of Vienna, Institute of Sociology

Aristotle (n.d., trans. 1984). Nicomachean ethics. In: Barnes J (ed) The complete works of Aristotle. The revised Oxford translation, vol 2. Princeton University Press, Princeton, NJ

Baddeley AD (1990) Human memory: theory and practice. Allyn & Bacon, Needham Heights, MA

Bertelsen OW, Petersen MG, Pold S (eds) (2004) Aesthetic approaches to human-computer interaction. In: Proceedings of the NordiCHI 2004 workshop, tampere, Finland, 24 October. University of Aarhus, Aarhus

Buttimer A (1980) Home, reach and the sense of place. In: Buttimer A, Seamon D (eds) The human experience of space and place. Croom Helm Ltd., London, pp 166–187

Chrysikou E, Tziraki C, Buhalis D (2018) Architectural hybrids for living across the lifespan: lessons from dementia. Serv Ind J 38(1–2):4–26

Colombo S, Djajadiningrat T, Rampino L (2015) Tangible, smart and dynamic objects. In: Chen L et al (eds) Design and semantics of form and movement. Eindhoven University of Technology, Eindhoven

Dahlin-Ivanoff S, Haak M, Fänge A, Iwarsson S (2007) The multiple meaning of home as experienced by very old Swedish people. Scand J Occup Ther 14(1)25:32

Desmet PMA (2003) Measuring emotion; development and application of an instrument to measure emotional responses to products. In: Blythe MA, Monk FA, Overbeeke K, Wright PC (eds) Funology: from usability to enjoyment. Kluwer Academic Publishers, Dordrecht, pp 111–123

Desmet PMA, Hekkert P (2002) The basis of product emotions. In: Green W, Jordan P (eds) Pleasure with products, beyond usability. Taylor & Francis, London, pp 60–68

Diener E, Eunkook MS (eds) (2000) Culture and subjective well-being. MIT Press, Cambridge, MA

Docampo Rama M (2001) Technology generations—handling complex user interfaces. University of Eindhoven, Eindhoven

Fung V, Lehmberg LJ (2016) Music for life: music participation and quality of life of senior citizens. Oxford University Press, New York

Gillsjö C, Schwartz-Barcott D, von Post I (2011) Home: the place the older adult cannot imagine living without. BMC Geriatr 11:10

Green WS, Jordan PW (eds) (2002) Pleasure with products: beyond usability. Taylor & Francis, London

Hassenzahl M (2001) The effect of perceived hedonic quality on product appealing-ness. Int J Hum-Comput-Interact 13(4):481–499

Hassenzahl M (2004) The interplay of beauty, goodness and usability in interactive products. Hum-Comput Interact 19:319–349

Hassenzahl M, Tractinsky N (2006) User experience—a research agenda. Behav Inf Technol 25(2):91–97

Hayward DG (1975) Home as an environmental and psychological concept. Landscape 20:2–9

Holman RH (1986) Advertising and emotionality. In: Peterson RA, Hoyer WD, Wilson WR (eds) The role of affect in consumer behaviour. Lexington Books, Lexington, MA, pp 119–140

Jordan PW (1998) An introduction to usability. Taylor & Francis, London

Jordan PW (2000) Designing pleasurable products: an introduction to the new human factors. Taylor & Francis, London

Leikas J (2009) Life-based design—a holistic approach to designing human-technology interaction. VTT Publications 726

Madden R (1999) Home-town anthropology . Aust J Anthropol 10(3):259–270

Maller C, Townsend M, Pryor A, Brown P, St Leger L (2006) Healthy nature healthy people: 'contact with nature' as an upstream health promotion intervention for populations. Health Promotion Int 21(1):45–54

Mallet S (2004) Understanding home: a critical review of the literature. Sociol Rev 52(1):62–89

McKay D, Cunningham SJ, Thomson K (2006) Exploring the user experience through collage. In: Proceedings of the 7th ACM SIGCHI New Zealand chapter's international conference on computer-human interaction: design centered HCI, pp 109–115

McMullin JA, Duerden Comeau T, Jovic E (2007) Generational affinities and discourses of difference: a case study of highly skilled information technology workers. Brit J Sociol 58(2)297:316

Medaesthetics. (2014) Creative ideas for art and design remedies in modern health care environments. Dementia village De Hogeweyk. 23 Feb. https://medaesthetics.wordpress.com/2014/02/23/dementia-village-de-hogeweyk/

Moloney MF (1997) The meanings of home in the stories of older women. West J Nurs Res 19(2):166–176

Monk AF, Lelos K (2007) Changing only the aesthetic features of a domestic product can affect its apparent usability. In: Venkatesh A, Gonzalvez T, Monk A, Buckner B (eds) Home informatics and telematics: ICT for the next billion. In: Proceedings of HOIT 2007, Chennai, India. Springer, New York, pp 221–234

Montague M (1999) Integrating the PRODUCT+BRAND experience. Des Manage J 17–23

Nagamachi M (1995) Kansei engineering: a new ergonomic consumer-oriented technology for product development. Int J Ind Ergon 15(1):3–11

Norman D (2004) Emotional design: why we love or hate everyday things. Basic Books, New York

Nussbaum M, Sen A (eds) (1993) The quality of life. Oxford University Press, Oxford

Oatley K, Jenkins JM (1996) Understanding emotions. Blackwell Publishers, Oxford

O'Bryant S (1983) The subjective value of 'home' to older homeowners. J Hous Elderly 1(1):29–43

O'Connor IJ (1997) Using attitudinal segmentation to target the consumer. In: Kahle LR, Chiagouris L (eds) Values, lifestyles and psychographics. Lawrence Erlbaum, Hillsdale NJ, pp 231–246

Osgood C, May W, Miron M (1975) Cross-cultural universals of affective meaning. University of Illinois Press, Chicago

Paulsson G, Paulsson N (1957) Tingens bruk och prägel (Things for everyday use and life form). Kooperativa förbundets förlag, Stockholm

Redström J (2001) Designing everyday computational things. Doctoral dissertation. Göteborg University, Gothenburg

Rioux L (2005) The well-being of aging people living in their own homes. J Environ Psychol 25(2):231–243

Rouaud J (1990/1998) Fields of glory (Trans. Ralph Manheim). Harvill Press, London

Rousi R, Saariluoma P, Leikas J (2010) Mental contents in user experience. In: Proceedings of MSE2010 V.II 2010 international conference on management and engineering, 17–18 Oct, Wuhan, China. ETP Engineering Press, Hong Kong, pp 204–06

Saariluoma P (2003) Apperception, content-based psychology and design. In: Lindeman U (ed) Human behaviour in design. Springer, Berlin, pp 72–78

Saariluoma P, Cañas JJ, Leikas J (2016) Designing for life—a human perspective on technology development. Palgrave MacMillan, London

Saariluoma P, Jokinen JPP, Kuuva S, Leikas J (2013) User experience as mental contents. In: 10th European academy of design conference—crafting the future. Gothenburg, Sweden

Saunders P (1989) The meaning of 'home' in contemporary English culture. Hous Stud 4(3):177–192

Scanlon T (1993) Value, desire, and quality of life. In: Nussbaum M, Sen A (eds) The quality of life. Oxford University Press, Oxford, pp 185–200

Schacter DL, Chiu CYP, Ochsner KN (1993) Implicit memory: a selective review. Annu Rev Neurosci 16:159–182

Schildt G (ed) (1997) Alvar Aalto in his own words. Otava, Helsinki

Schütte S (2005) Engineering emotional values in product design—Kansei engineering in development. Dissertation No. 95. Institute of Technology, Linköping

Sixsmith JC (1986) The meaning of home: an exploratory study of environmental experience. J Environ Psychol 6(4):281–298

Smith SM, Vela E (2001) Environmental context-dependent memory: a review and meta-analysis. Psychnomic Bull Rev 8(2):203–220

Solomon RC (1993) The philosophy of emotions. In: Lewis M, Haviland JM (eds) Handbook of emotions. The Guilford Press, New York, pp 3–15

Steg L, Van den Berg AE, De Groot JIM (2012) Environmental psychology: an introduction. Wiley-Blackwell, London

Swenson MM (1998) The meaning of home to five elderly women. Health Care Women Int 19(5):381–393

Tomkins SS (1984) Affect theory. In: Scherer KP, Ekman P (eds) Approaches to emotion. Lawrence Erlbaum, Hillsdale, NJ, pp 163–195

Tractinsky N, Katz AS, Ikar D (2000) What is beautiful is usable. Interact Comput 13(2):127–145

Veenhoven R (2012) Happiness: also known as 'life satisfaction' and 'subjective wellbeing.' In: Land KC, Michalos AC, Sirgy MJ (eds) Handbook of social indicators and quality of life research. Springer Publishers, Dordrecht, Netherlands, pp 63–77

World Health Organization (1994) Statement developed by WHO quality of life working group. In: WHO health promoting glossary 1998. WHO/HPR//HEP/98.1. World Health Organization, Geneva

Williams AM (2002) Changing geographies of care: employing the concept of therapeutic landscapes as a framework in examining home space. Soc Sci Med 55(1):141–154

Williams AM (2004) Shaping the practice of home care: critical case studies of the significance of the meaning of home. Int J Palliat Nurs 10(7):333–342

Wright P, McCarthy J, Meekison L (2004) Making sense of experience. In: Blythe MA, Overbeeke K, Monk AF, Wright PC (eds) Funology: from usability to enjoyment. Kluwer Academic Publishers, Norwell, MA, pp 43–53

Ziefle M, Schaar AK (2017) Technology acceptance by patients: empowerment and stigma. In: Van Hoost J, Demiris G, Wouters E (eds) Handbook of smart homes, health care and well-being. Springer, Berlin, pp 167–77

Chapter 11
Emotions and Technoethics

Pertti Saariluoma and Rebekah Rousi

Abstract The relationship between emotions and ethics has been debated for centuries. The act of understanding emotions through the framework of ethics involves accepting that emotions are to some extent culturally dependent. By linking emotions in design to larger ethical discussions, it may be accepted that ethics and design are both technological constructions designed to shape a collective world-view. While both are cultural constructions, they are in constant dialogue with one another through social discourse and individualistic cognitive–affective appraisal processes. This chapter presents an account of technoethics that challenges ideas of ethical values embedded within technology, drawing attention to the role of human intentionality as a definitive ethical factor in human–technology relationships. The chapter problematises simplistic views of ethics and emotional technology experience to reveal the ambiguous and dynamic nature of cognitive–emotional–cultural interdependencies in technology experience.

11.1 Introduction

At the heart of all technology is the intention to make human life easier in some way. Technology speeds up processes to make them more efficient (and often more effective) and relieves people's physical and mental burdens. It can enable the faster and more accurate delivery of completed tasks and give individuals the opportunity to accomplish actions that would never have been possible otherwise. For instance, steam engines not only ran *according to* timetables and time zones; they *enabled* people to take loads on time from one place to another (see, e.g., Zerubavel 1982). Similarly, combustion engines rendered air travel possible (Hiereth and Prenninger

P. Saariluoma (✉)
University of Jyväskylä, Jyväskylä, Finland
e-mail: pertti.saariluoma@jyu.fi

R. Rousi
University of Jyväskylä and Gofore, Jyväskylä, Finland
e-mail: rebekah.rousi@jyu.fi

© Springer Nature Switzerland AG 2020   167
R. Rousi et al. (eds.), *Emotions in Technology Design: From Experience to Ethics*, Human–Computer Interaction Series,
https://doi.org/10.1007/978-3-030-53483-7_11

2007), which exponentially increased the pace of mobility and communication. These forms of technology changed human life permanently and dramatically. Humans learnt to live in radically different ways. This same rule may indeed apply to the COVID-19 era.

From axes to software applications, technical artefacts are tools people utilise to pursue their objectives—life goals as well as mundane everyday tasks. Yet, goals and actions have ethical dimensions and consequences. The intention behind the action is the definitive factor that determines whether the goal is 'good or evil' (Ferrarello 2015; Shotter 1995). In other words, do the goals correspond to ethics and codes of moral conduct—i.e. how people have learnt to behave in the correct way according to cultural principles. An additional consideration is whether, by behaving in a certain way, is the actor treating others how they would want to be treated (Gensler 2013)? Or is the intention behind the goal and associated actions to generate gain at the expense of others? Actions associated with these goals may be deemed necessary, allowed or forbidden (von Wright 1963). Yet, the relationship between these dimensions may be (and often is) extremely complex. The concepts and rules of ethics are intimately connected to the use of technical artefacts. Thus, at any stage of its lifespan, technology should be considered in terms of its ethical aspects. Mario Bunge coined the term 'technoethics' in 1974 to emphasise the ethical responsibilities of technologists (Bunge 1977). It is increasingly used to describe the concurrent relationship between technology and ethics and how they exist in relation to moral codes.

Questions of ethics have been debated for centuries. Ethics by nature trigger emotions, both in relation to the topic itself and the framework it provides for evaluating phenomena. Moreover, ethics incite emotional reactions within people regarding their own actions and the consequences of these actions. Hume (1751/1998), like many other British, American and Commonwealth philosophers since, directly linked ethical actions to human emotions (Ayer 1936; Hume 1751/1998; Moore 1991; Stevenson 1944). As Hume (1888, 457) explained, 'Morals excite passions, and produce or prevent actions. Reason itself is utterly impotent in this particular. The rules of morality, therefore, are not conclusions of our reason'.

Hume's position connects ethics with emotions through the control of actions. The act of understanding ethics through emotions is referred to as 'emotivism' (Malik 2014; van Roojen 2018). According to emotivism, ethical propositions—and thoughts—are manifested and expressed through emotional states rather than cognitively produced and explicit facts (Ayer 1936; Hume 1751/1998; Stevenson 1944). Therefore, emotions and emotional states seem to play a crucial role in defining what is ethical in the human mind. If the consequences of actions lead to states that are emotionally negative or potentially destructive and harmful, it is often assumed that these actions should not be undertaken. If an individual cognitively processes (thinks about) these cause–effect relationships, they may decide not to take such actions. Performing such actions regardless of the consequences causes people to enter a state of cognitive dissonance (Stone and Cooper 2001). This also can be classified as moral dissonance—when an individual acts against his or her moral codes and beliefs (Breslavs 2013), inducing high levels of negative arousal (intensely experienced and unsettling negative emotions). The dynamics within this state are highly complex and produce an understanding of emotions that is both multidimensional and conflicting.

For instance, an educated and aware individual may know that the development of artificial intelligent (AI) systems is based on exploiting vast amounts of data collected from people without their knowledge. Whether individuals have given their consent through pop-up messages that interrupt their initial web page viewing without carefully reading the privacy disclosures or understanding the nature of 'cookies', or whether the data utilised have been obtained entirely without individuals' consent, substantial amounts of data (big data) are needed to 'feed' machine learning. In this example, the educated individual in question comprehends that from one perspective, AI developers are in fact stealing data from other individuals in order to enable the system to learn and operate. Thus, a negative emotional tone is set regarding the ethical correctness of the technology. However, the AI system is convenient, efficient and enables the individual to perform complex actions that they would not otherwise be able to do. This helps them to fulfil their career and life goals. The dissonance stems from the fact that the individual is not entirely emotionally positive (happy) with the ethics behind the technology's development but they are indeed satisfied with how it works as an enabler. In sum, the essence of emotivism is the association of ethics and moral codes with the human emotional system and its properties.

Technology, human actions, culture ethics and emotions are all tightly interwoven (Chen et al. 2020; Hume 1751/1998). Where culture products—including technology—are concerned, ethics always provide a framework, whether explicit or implicit, for the complex emotional reactions that individuals experience in relation to technology design and its consequences. Thus, it is relevant to discuss the relationship between emotions and technoethics, as within the context of technology design experience (from the perspective of both the design team and users), the two cannot be neatly separated. This chapter focuses on the conceptual foundations of studying emotions and technoethical synergies. It emphasises that ethics can be understood as one of the main links between emotions and technology. As the above example illustrates, along with the nature and rapid pace of the technological development of today's world (and arguably throughout human history), ethical values have evolved.

Emerging AI systems dominate the current cultural landscape. Robots, chatbots, other bots, agents, AI in social media (e.g. Facebook's Sophia)[1] and autonomous cars will change the way people live. Transportation, industrial banking, culture, administration, medical care and learning are already experiencing dramatic shifts (Tegmark 2017). These shifts involve processes and operational models as much as they entail the rupturing and reformation of human–technology and human–human relationships. In an instant, the COVID-19 crisis radically altered human–human

---

[1]Facebook's controversial 'Sophia' (see https://www.facebook.com/realsophiarobot/) is one of the social media's first public campaigns to integrate AI and seek public acceptance. Controversy regarding its development ranges from the role of personal data in its development to accounts of Facebook 'staging' Sophia's intelligence in a Wizard-of-Oz (puppetry) type manner.

interactions. Learning, meetings, business, clinical and diagnostic sessions (health and mental care and other therapy/life exchanges), cultural consumption (concerts, theatre, art, etc.) and research suddenly moved from face-to-face events to internet-mediated transactions. When it was possible to meet in person, many chose to meet other individuals in an unmediated fashion due to their own moral stance—what they considered as being ethically correct interpersonal behaviour.[2] Traditional academic discourse and public sentiment have frequently mentioned concerns for technology (computers, robots and AI) in terms of compensating for genuine human-to-human contact (see, e.g., Barnes 1996; Cerulo 2009; MacDorman et al. 2009; Turkle 2007). Yet, during an epidemic, there is a simultaneous need to maintain distance while also remaining connected. Thus, many have embraced information technology for its ability to uphold social life and real-time interactions—even those who previously found it ethically challenging.

Every technological innovation generates effects that can be experienced to varying degrees across society. For instance, the development of agriculture has had fundamental repercussions across all areas of human life—technological, behavioural, societal, psychological, etc. (Bernal 1969; Hendrick 2009)—which have inevitably shaped current human societies and cultures. Another example is gunpowder, which can be understood as the bridge between the medieval and modern eras of human societies (Britannica 2020). Its invention was as much about how it was created as it was about what it did. The development of gunpowder marked the first true application of theory (reason and rationality) to empirical experimentation in harnessing and exploiting energy. This, in turn, laid the foundations for scientific problem-solving and development through equations and theoretical modelling before it was applied in real-life settings. Chain reactions from this discovery can be seen across society, from scientific and educational institution formation and curricula development to implementation of the technological breakthroughs that gunpowder-related discoveries enabled (e.g. the ballistic pendulum for velocity measurement). Yet in this case, the ethical basis and conflict—dissonance—between moral principles (taking another person's life) and technosocietal advancement constitute an uneasy relationship between emotions and technoethical systems.

11.2   Ethical Neutrality of Technical Artefacts

Without knowledge of concrete uses, cultural framing or understanding of the design intention, technical artefacts and objects themselves are ethically neutral. When encountering a device or technical artefact, without background knowledge, it is

---

[2]It could be argued that technology always mediates human interactions, for example, via computer systems, in the context of architecture (classrooms, schools, homes, etc.) or commerce (cafes, restaurants, shopping malls). Yet in this instance, we are referring to the unfiltered, direct means of communication from one individual to another.

generally impossible to say whether it is good or evil (i.e. if it will be used for benefit or detriment). Technology is culturally constructed and exists as an expression of cultural values, beliefs and social ways of being and doing (societal systems). Underlying its development, there is always intention and intentionality—a knowingness and goal-related state of being that direct human actions and shapes human-generated products. However, there is no guarantee that even technical artefacts created with the best of intentions will not be used in unethical ways to reach unethical goals.

The relationship between what is understood to be either good or evil is multifaceted and complex and depends on a range of factors: (a) who is creating, (b) what is being created, (c) why it is being created, (d) for whom and (e) how others who are not involved in the development will receive the results of the production. In short, the techoethical relationship between a person and the design depends entirely on the relationship between the producer(s) (the commissioner(s) and the creator(s)) and the receiver(s)—customers, users, citizens, allies and opponents (Hodgson 1983). If an individual has no prior knowledge of an object when it is encountered for the first time, it appears to be simply a mass and form of materials (Ramsey 2016), even if the technology has been constructed or composed intentionally (Borgmann 2012; Cooley 1995; Jonas 1982). Intentionality still rests within the minds of humans within human–technology relationships (Cardon 2018; Devillers 2020; Haikonen 2020). The technology itself is incapable of intentionality or consciousness. Therefore, the human–technology relationship defines technoethical ambiguity.

This ambiguity between an artefact and its uses can be illustrated using the example of drones. A drone can be used for a range of humanitarian acts such as transporting food and medicines as well as aiding the performance of actions in inhumane circumstances such as warfare and unsolicited surveillance. Thus, drones are also notorious for their invasive qualities; governments worldwide have established laws forbidding their flight within certain distances of private houses due to their potential use for spying and stalking and taking unauthorised photographs. Drones are also associated with drugs and weapon smuggling as well as unfair, unethical and unprecedented warfare activity against civilians in various countries. This mixture of humane and inhumane intentions, and the use and actions associated with drone technology, renders emotional sentiment towards them conflicting and controversial. Yet a drone cannot be rendered either good or evil in its own right. Only when it is attached to human intention, behaviour and cultural framing—the discourse surrounding the technology (in the media, political, news, entertainment, etc.)—do it enter a conceptual dialogue with the individual and group's ethical basis.

The term 'technology' has multiple meanings and connotations. It refers to the practical application of theoretical knowledge (Merriam-Webster 2020); its history was discussed above in the development of gunpowder. From a historical perspective (derived from the Greek word *techne*), technology is anything human made; it is closely connected to understandings of tools and other artefacts humans use to shape their environment (Derry and Williams 1960). What is now understood as art played a major role in the formation of history, collective memory and the infusion of human imaginings into the surrounding physical environment. Ethical frameworks and networks of trust are established through action and speech or speech acts

(McNair and Paretti 2010; Jarvenpaa et al. 2004). The establishment or disbandment of trust and ethical evaluation processes may be triggered by something as subtle as an utterance—e.g. 'Be careful, online games can be dangerous'—a locutionary act (Austin 1975) or framed by a more conventional political speech act (locutionary act).

While Bauhaus and Volkswagen have attained relatively widespread acceptance on international markets, these movements and brands were associated with the Nazi regime through Adolf Hitler's political rhetoric in *Mein Kampf*[3] (McGuire 1977; O'Shaughnessy 2009). Without taking into account, the cultural–political context, Bauhaus design may be understood from a functional perspective as furniture. Volkswagen could likewise be understood simply as a car manufacturer—on a basic level, producing vehicles for transportation. Yet, through deeper associative thinking and knowledge of political world history, the technology and design produced by both brands can be connected to Nazism and crimes against humanity, disrupting the moral grounding for the consumption of their products. The design objects are no longer simply objects but tools of Nazi propaganda and political rhetoric connected to the Holocaust. From a functional perspective, cars can be understood as enabling faster, broader travel opportunities than horses and trains. While seemingly purely functional, from another ethical perspective, Volkswagen vehicles and automobiles in general may be seen as speeding the pace of human communication and travel for the purpose of exponentially increasing productivity. Thus, an ethical evaluation of cars may explore 'Who *really* benefits from the increase in human speed?' There is increased productivity and increased consumption, all at the price of the increased risk of serious accidents and potential fatalities.

Thus, on the surface, objects may appear to be technoethically neutral. Yet, the act of creating (designing and producing), the act of using and discursive (speech) acts surrounding the technology design always ricochet continuously across a spectrum of ethical alternatives. These in turn affect the ways in which people emotionally experience technologies and their uses. Ethical issues that are relevant in acting with technical artefacts should always be considered in the context of people and how they use technology in their physical actions and speech. AI is no exception. Its ethical value is based on the ways in which people design, implement and utilise AI capabilities. This means that in order to consider emotional weighting and experience during the design and development process, design teams should chart and account for both possible uses associated with intention—pro-humane versus inhumane (for others' benefit or to others' detriment)—and possible interpretations (i.e. connections to history, culture and socio-political discourse).

---

[3]Nicholas Salas (2013) provides an interesting account of the development of Volkswagen from Hitler to the US market. See https://thevisualcommunicationguy.com/2013/07/03/how-adolf-hit lers-ideal-car-became-an-american-favorite-a-visual-analysis-of-the-volkswagon-beetle/.

11.3 Actions, Rules and Ethics

Human actions are based on cognitive information processes. People are information-processing animals; thus, all aspects of human actions should be considered information processes and the results of human information processing (Newell and Simon 1972; Norman 1976; Rumelhart et al. 1972). This section describes technoethics in light of information processing and actions guided by information processes.

Individuals and groups are guided by ethical norms and principles. Ethical consciousness, or ethical and moral sensitivity, varies between people (Knapp et al. 2009). While every individual has a moral code as well as implicit and explicit sets of ethical principles, some people more actively seek information and engage in ethical reflection than others (Cohen et al. 2001). Ethically conscious people try to avoid engaging in situations and making decisions in which other people, or groups of people, are placed at a disadvantage. Thus, when adhering to ethical codes such as the Golden Rule—the Code of Hammurabi—individuals understand that they should treat others how they would like to be treated themselves.[4] The phrase 'an eye for an eye' originated from this code. It is claimed that this code, and Hammurabi himself, is accounted for in the Hebrew Old Testament in the story of Moses (the Mosaic Code) in which God (Yahweh) inscribed a set of laws, the Ten Commandments, on two stone tablets. These laws are: (1) worship only one God, the genuine *God* and no other (interpreted in many ways, this can be seen to emphasise the importance of spiritual awareness and humanity and not to prioritise anything else); (2) do not disrespect God; (3) keep a day aside each week for God and spiritual health (rest); (4) respect one's parents; (5) do not murder; (6) do not commit adultery; (7) do not steal; (8) do not lie; (9) do not lust after someone else's partner and (10) do not be greedy or jealous (avoid ill intentions). Thus, in many ways, the Ten Commandments (the code) provide the foundations for modern-day law (Green 2000).

In both Western and non-Western societies, the premise of treating others how one would want to be treated is a benchmark for social–cultural being (civilisation) and ethics (United Nations 2020). Respect (Jing) is central to Confucianism (Chan 2006), tied to this are the elements of religion, moral rules and etiquette. The Confucian Golden Rule consists of two parts, *chung* and *shu* (Ivanhoe 1990). These are referred to as the 'One Thread' of the *Analects of Confucius* (a book of sayings) as they are tightly woven together (Ivanhoe 1990, p. 17), representing reversibility—the notion that actions taken on other people are subsequently returned to or reversed back upon the actor (Fung 1953). *Chung* represents the things that one should do for others, and *shu* represents what should not be done to others (Ivanhoe 1990). In Islam, it is stated that '*None of you believes until he wishes for his brother what he wishes for himself*' (An-Nawawi's Forty Hadith 13, cited in Islamic Network Group 2020).

---

[4]Hammurabi was the sixth king of the first Babylon dynasty (Prince, 1904). He was a god-fearing yet powerful soldier. He ensured his people's well-being by developing an infrastructure and societal order via rules and laws. These laws instilled an understanding among his people that they should treat others how they would want to be treated.

All religions promote the Golden Rule in some way; thus all cultures have more or less the same basic underlying framework for ethics and moral principles that shapes people's ways of relating to one another. Thus, ethics and the information attached to them—the knowingness or awareness of when people act towards others in a positive or negative (exploitative) way—are to varying degrees so embedded within the social–cultural being of humanity that they are intrinsically connected to emotions. This is particularly true for higher-order cognitive processing, in which deeper associative thinking—the information processing that is connected to evaluating, or appraising, cultural products (technology)—connects an individual to the reflective state of what the design, its creation, production, and consumption means for oneself and for others.

Yet, like any other area of emotional scholarship and human experience, the relationship between ethics and emotions is complex. Ethical information processes can be studied from various points of view. Ethical norms and changes in moral codes can be seen as a result of alterations in ethical and legal discourses. While laws and regulations have been influenced by the Golden Rule as discussed above, they are still cultural constructions in which people have actively created social and behavioural frameworks by which citizens must abide (Dror 1957). Thus, depending on legislative systems and governmental structures, measures may be taken to establish laws that somewhat go against basic human values as set forth in the Golden Rule, possibly creating major dissonance in the beginning, yet gradually smoothing out as the codes of behaviour become normalised and accustomed to. From a historical perspective, these types of laws can be seen in relation to Nazi Germany's treatment of Jews and South Africa's apartheid.

In his *Critique of Practical Reason* (1788/2004), Immanuel Kant described a rational approach to ethics. He argued that stealing, murdering, deceiving and adultery, for instance, would spell the end of civilisation. In order to bypass this form of reasoning, political entities have institutionalised a form of *othering* that exceptionalises other cultural and ethnic groups, rendering them outside the realm of human values and civilisation. The indigenous people of Australia, for example, were not recognised as having cultures of their own due to a lack of recognisable European political structures and other technological designs—i.e. architecture and post-enlightenment cultural and institutional establishments. This meant that they were not recognised as Australian citizens until 1948 (1944 in Western Australia) and rendered them unable to vote in parliamentary elections until 1967 (Western Australian Museum 2017). Institutional and individual acts of racism (breaking the Golden Rule) were somewhat accepted, as indigenous people were not considered 'neighbours' or 'siblings', but rather as beings that existed outside of culture and civilisation—not adhering to the Golden Rule and resembling animals more than humans. To institutionalise and thus cultivate this understanding, indigenous Australian artefacts and remains were placed in the technology of museums (text, display, architecture, layout and institutional discourse) in Australia and throughout Europe—separating them from what was considered the humanised norm of European society (Giles 2006; Turnbull 2007).

This act of ethical value shifting can be seen across the board of institutionalised humanity and is one of the main mechanisms of war. Culture—art and technological design—has played instrumental roles in reframing the relationship between humans and ethical standards, as witnessed in the tight connections between Hitler, Bauhaus and Volkswagen as well as Filippo Marinetti and the Futurist Movement with Benito Mussolini. Fascist ideas shaped human life, behaviour and society at the political governance level *and* at the practical life-shaping technological design level. Modern understandings of the ways in which technology, business and public discourse also shift ethical values can be seen in the acceptance of apps such as Tinder, Ashley Maddison, Nude and even Facebook—in which promiscuity, deception and adultery are simply part of the design and business model. Axiology (the study of the nature of values) is extremely useful in this context, as the shifting of ethical values and manipulation of the Golden Rule may be a sign of greater questions or challenges facing humankind.

Ethical values and corresponding actions may be considered from an empirical perspective, particularly in relation to the ways in which people's behaviour changes according to the context. This is no modern insight; for example, people may behave radically against the Golden Rule and then seek absolution upon entering a designated site of spirituality (the confession box). Alternatively, in line with Kant (1788/2004) and his ideas about how people should act in an optimal way that one would hope other rational people would act—just as if it were a universal law—ethics could be considered contextually optimal psychological and social processes. This helps explain why there are subtle variations in interpretations of the Golden Rule between cultures, as no two cultural groups have exactly the same conditions (Westermarck 2017).

Ethical information processes and their analyses represent a specific approach to the study of ethics, which can be supported by its importance in designing an optimal world according to the ethical values. To use an illustration from elderly care, instead of simply representing external academic norms related to the *right*, or correct, kind of patient care, design teams can strive to understand how people are really cared for, for example, in units for senior citizens and what norms caretakers follow in their daily lives. This type of empirical ethics is intimately connected to the analysis of ethical processes but with an important difference. The former moves the focus from academic discussions to life as people live it, which leads to the tacit and explicit development of a society's ethical framework and moral codes. The latter refers to the analysis of how norms are created, serving as an empirical model of meta-ethical processes in everyday life.

In the late 1800s and early 1900s, Westermarck (2017) studied the relativity norms and values of empirical ethics. His analysis of ethical information processes takes a slightly different form: it concentrates on the process of creating the social norms and ethical values people follow in their daily lives. Drawing on the traditions of John Stuart Mill, Westermarck's approach was against the normalisation of ethics and any objective understanding of morals. Instead, his work focused on examining the contextualisation and situational relatedness of ethics and moral codes. That is, Westermarck argued that depending on the situation in which an individual found herself, her ethical stance should adjust according to what would be most useful. Thus, utilitarian ethics are guiding rules that maintain individuals should act in a way that would instil the greatest amount of happiness for as many people as possible. As Mill (1863) stated:

> Now the utilitarian standard is not the agent's own greatest happiness, but the greatest amount of happiness altogether. It may be defined as the rules and precepts for human conduct by the observance of which happiness might be, to the greatest extent possible, secured to all mankind…

Both Westermarck and Mill go beyond utilitarianism to argue that ethics also form the moral code of behaviour that supports people's desire to be connected to one another. Human beings harbour social feelings towards their fellow humans, which increases their personal interest in ensuring the well-being of others. To this end, the creation of values and how these are adopted and abided by as social processes are important in research on ethical information processes. To contextualise this ethical discussion within contemporary technology design, development and business, ethics can be seen here to be based on the analysis of real-life value creation processes, or 'process ethics', in order to distinguish this way of understanding ethics from more static and normative approaches.

Value creation is an important element of design thinking. Design is a value generation process. If researchers understand the value creation process in light of the ways in which ethical values exist, operate and morph, they can improve the progress of design by providing empirical information from various angles of the process. This shift from reflective to active involvement and influence is vital when designing ethical AI systems. Academic discussion is one example of a value generation process, but administrative, journalistic and law-making processes are equally important for the overall framework of how ethics and emotions influence one another in technology design. Moreover, the most important value generation process, the arena within which the ethical guidelines and moral codes are discursively and socially formalised in an open society, is nevertheless within popular discourse and discussions between citizens.

11.4   A Glimpse at the Axiology of Technoethics

Ethics are social, cultural and psychological mechanisms that regulate how people should act. Knowledge of what is allowed or forbidden is expressed in the form of ethical principles and rules. Today, there are ethical requirements for AI and other emerging technologies. This knowledge opens an axiological or rule-based view to future technoethics. Ethical rules can be divided into two broad classes. The first is fundamental rules, which have been widely accepted over millennia such as the Golden Rule. The second can be seen in the forms of ethical guidelines. These have recently emerged, for example, to clarify how people should treat one another in online spaces. The Association of Internet Researchers (2020), for instance, has published four sets of guidelines since 2002 in an attempt to prevent researchers from exploiting individuals or private data on the internet. The guidelines state that all internet users are responsible for treating all content providers as vulnerable individuals whose privacy should be maintained and intellectual output respected.

Axiology (value theory) is the study and structural theory of values (Bahm 1993; Hart 1971; Hartman 2011). Axiology can be applied to study the formation of ethical rules and has been used quite extensively in areas such as environmental ethics (Muraca 2011). It has its origins in Moore's (1959) *Principia Ethica.* Findlay (1970) noted that axiology was treated as the 'tail end of ethics'—values being created through ethics. He highlighted this as an irony, since axiology should be considered in light of the *formation* of ethics. Axiology was originally discussed in terms of the study of 'ultimately worthwhile things' (Findlay 1970, p. I). Yet, even according to this early definition, it is difficult, if not redundant, to argue about which one is more important. This chapter has so far detailed that ethics are formed out of interest (value) in humanity and well-being for fellow humans; values can be formed through ethics, yet ethics also shape these values and the level of human sensitivity (ethical sensitivity and awareness) to them. Ethical rules represent and promote the maintenance of human values; they explicate how people should act or live in an ethically correct way. In Plato's terms, such rules aspire to ensure maximum happiness (Frede 2017). In technoethics, ethical rules express how people should utilise technological capacity to support their quality of life. This means that from both design and development perspectives as well as the usage perspective, technology should be developed and used for the benefit, well-being and *happiness* of as many people as possible. This line of ethical thinking should guide people in their decisions and subsequent actions in any technologically related situation.

Typical examples of ethical principles can be found, for example, in the UN's Universal Declaration of Human Rights:

> A1: All human beings are born free and equal in dignity and rights. They are endowed with reason and conscience and should act towards one another in a spirit of brotherhood.
>
> A3. Everyone has the right to life, liberty and security of person.
>
> A26.1: Everyone has the right to education. Education shall be free, at least in the elementary and fundamental stages. Elementary education shall be compulsory. Technical and professional education shall be made generally available and higher education shall be equally accessible to all on the basis of merit.
>
> A26.2: Education shall be directed to the full development of the human personality and to the strengthening of respect for human rights and fundamental freedoms. It shall promote understanding, tolerance and friendship among all nations, racial or religious groups and shall further the activities of the United Nations for the maintenance of peace.
>
> A26.3: Parents have a prior right to choose the kind of education that shall be given to their children.

These examples of human rights are very general, but they can be used to derive ethical principles for technologies. For example, as with the Golden Rule, the first article suggests that one should not use technology to harm other people—particularly their liberty, life or security. Article 26 encourages the use of technology for egalitarian purposes, affording equal access to education.

Examples of technoethical rules specify general ethical principles for technology, and especially emerging AI systems may be for instance:

E1: Individuals should be respected, and technological system solutions should not violate their dignity as human beings, or their rights, freedoms and cultural diversity.

E2: Individual freedom and choice. Users should have the ability to control, cope with and make personal decisions about how to live on a day-to-day basis, according to one's own rules and preferences, rather than to be forced into technologically oriented ways of doing and thinking.

E3: Concerns the role of people and the capability of AIS to answer for the decisions and to identify errors or unexpected results. AIS should be designed, so that their effects align with a plurality of fundamental human values and rights, this means taking into account diversity between groups and individuals.

Example 1 promotes human values through respect for individual freedom and diversity. This is similar to Example 2, in which people should additionally have the ability to control and make personal decisions regarding the technology they are affiliated with, in order to adopt, adapt and appropriate the designs to their own life circumstances. As is observed in Example 3, technoethical rules specify uses of advanced intelligent systems such as AI to assist in the maintenance of human values and rights. These examples lead to the development of an overall image of ethical human–technology relations that nurture human dignity. Human dignity should in turn provide the basis for determining the direction, ethical guidelines and governance of technology and its design (Zardiashvili and Fosch-Villaronga 2020). Ethics apply to all people; thus ethical principles should be applied to both the design and utilisation of technological systems. To more clearly understand the relationship between content associated with ethics and connections to emotional aspects of technology, the next section explores how these ethical rules are created in social actions.

11.5   Technoethical Process

Hume (1751/1998) discussed the role of emotions in creating ethical rules. He argued that value does not simply emerge from facts. Reason operates via descriptive propositions that cannot control the emotions that are essential elements in prescriptive, ethical propositions. For instance, knowledge of the dangers of particular substances does not automatically lead to the idea that one should not drink or smoke. In other words, despite knowing the dangers of alcohol and tobacco, people do not necessarily have an automatic negative emotional reaction towards these substances. Instead, the alcohol and tobacco industries are alive and well due to the fact that the 'dangerous' element of knowledge about these substances is not prioritised. Other elements and associations—highly social—are connected with these such as partying, relaxing, socialising, etc. Thus, Hume's guillotine, or the *is–ought* (to be) argument, is still an important ethical dilemma and one cannot say that it has been solved (Saariluoma 2020). Moreover, this dilemma may always be present as it represents the conflict between shared human values—human rights and respect for others—and vested interests in oneself disconnected from the common good (Black 1964; Kazavin 2020). To gain clarity on this issue, the cognitive–affective act of processing ethical information must be analysed.

Human experience, i.e. conscious mental representation in the human mind, forms a central component of human information processing and thinking (Chalmers 1996; Dennett 1993; Rousi 2013a). The information content of experiences and mental representations can be called mental content (Saariluoma 2003). Mental representations have cognitive and emotional dimensions (Rousi et al. 2010). Both play an important role in ethical information processing. From one perspective, Hume's guillotine separates these: cognitive processing based on raw facts does not necessarily incite or become connected to emotional processing, yet other matters such as self-investment and/or social (socio-economic) benefit carry emotional weighting (value). Thus, the saying '*do as I say and not as I do*' represents the conflict between cognition, action and emotion. Moreover, from an ethical perspective, it is also important to understand the role of emotional valence in mental content.

Ethically, an important type of mental content is emotional valence (Oatley 1992). Although emotions and their processing and experience are complex, most emotions can be divided into those with either positive or negative associations. However, the so-called negative basic emotions such as sadness (e.g. grief, death of a loved one) or anger (fury towards the offender) may be interpreted and even experienced in a positive way. For instance, sadness about a loss may be linked to happiness for a friendship or anger about being betrayed may be seen as a motivation to heal and move on. Thus, valence and its felt intensity (arousal, see, i.e., Kron et al. 2015 and Robinson et al. 2004) are crucial aspects of experienced emotions and their cognitive functions. Once again, these reactions are generated based on complex, dynamic processes that inevitably—consciously and subconsciously—inform an individual whether or not the evaluated (appraised) phenomenon in question is beneficial or detrimental to an individual's personal well-being. This cognitive–affective process is referred to as appraisal (Folkman and Lazarus 1984; Frijda 1993; Lazarus and Smith 1988). Appraisal, and this complicated relationship between an individual's emotions and ethics, poses perhaps the greatest challenge to Hume's guillotine. While emotions may be influenced at the cognitive level by knowledge of ethics and the need to act for the maximum benefit of oneself and others, from the individual emotional perspective, opportunities may be observed through technological capacities such as the ability to collect mass amounts of data (big data) for the purposes of modelling, targeted marketing, sale (business intelligence) and further exploitation, which give the individual technologically and economically competitive edge at the expense of others. That is, individual gain is prioritised over collective gain.

Valence arguably renders emotivism and emotionally loaded ethical thinking possible as it tempers the semantic value of cognitive–affective content according to benefit or detriment and relational ways of being (Joseph 2009; Wolff 2019). Often linked to linguistics (the technology of language) and other cultural expression, emotivism in ethics begins with the idea that situations of life and respective experiences are either emotionally positive or negative (pleasant or unpleasant). The emotional analysis of the consequences of actions thus provides the basis for the ethical analysis of actions and action types. For example, the Golden Rule can be seen as a generalisation of situational experiences of deeds in which the principle is either followed or violated. Thus, emotional and ethical information processing involves the analysis and experience of emotional valence in terms of emotional meaning (the consequences for the individual as well as others or in light of others) and can be considered the first point of the ethical process.

Consequently, the development of ethical norms is grounded in the quality of valence that people associate with various emotional situations. However, it is not possible to finalise the analysis of the ethical process with emotions and their valences (Saariluoma 2020; Saariluoma and Leikas 2020). Life situations are the consequences of human actions—not necessarily carried out by the individual herself but by humans and their constructions in general. Thus, the value of actions can be defined based on the valence of situations that arise as a consequence of particular types of actions. Norms describe what kinds of actions have had emotionally positive or negative consequences. In many contexts, and as a discursive 'rule of thumb', actions leading to pain are not generally acceptable, while actions leading to positive emotions are often referred to as 'good'. Yet, in the context of punishment and war, pain inflicted on the perceived offender or enemy can be emotionally experienced as positive by a collective group of individuals. Likewise, positive emotions experienced while engaging with video games can be considered negative from the perspectives of game addiction, reduced social engagement, physiological problems and depression.

The first step in defining technoethical principles is to classify situations according to their emotional values. That is, what types of emotions (positive–negative; aroused–passive) arise in relation to technology and associated factors—context, intention, technological type, logic, etc. From an emotional perspective, actions leading to various technology interaction situations can be organised into two types of categories with a follow-up question—beneficial or detrimental and for whom (the individual, the collective, humankind in general)? Even in situations in which emotions are not consciously experienced, such as elevator travel or basic graph-ical user–interface interaction, an ongoing evaluation process is always occurring. In studies on elevator use, for instance, Rebekah Rousi (2013b; 2014) observed that the 'experience of no experience'—no conscious emotional experience—revealed that the technology interaction had been positive. That is, the relationship between positive and negative technological experiences with conscious and unconscious emotions is complex, dynamic and highly contextualised (Winkielman and Berridge 2004). In the context of elevator use, an individual engages in a fully embodied interaction, meaning that the stakes are high in terms of safety and security. Faulty function in the technology would mean either serious injury or loss of life. Thus, negative interactions entail a conscious emotional experience.

At the interaction design level, issues relating to mass data collection, cookie use, clickbait and indeed AI and bots may be dismissed as minor. In other words, users may not think beyond the surface (interactional) level, whereby symbolic interactions matter (Saariluoma and Rousi 2015). These symbolic interactions may not generate strong conscious emotional reactions. Yet, upon further reflection, higher order cognitive emotions are engaged through consideration for reasons behind internet-based data collection and subsequently the further use of collected data (financial gain and social–political control). Consider the knowledge of personal data being sold onwards: the owners of the website generate profits from one's data without consent or offering royalties to the users. Then, consider the intentions of those who buy the data: Are they in line with one's own values, desires and interests related to personal safety security and freedom? For example, based on engaging in a seemingly free online game, what do the webpage owners gain at the eventual expense of the users? If these ethical questions—and, moreover, facts—were readily apparent to users, more effective and reactive emotional reactions would be experienced.[5]

The generation of ethical norms is a cultural constructive process that involves emotivism and allocating ethical–emotional meaning (Cohen et al. 1992; Hofstede 1980; Jeffrey et al. 2004; Markus and Kitayama 1994). Knowledge of the need to avoid alcohol use, for instance, as it leads to health and social problems may or may not serve as a deterrent for alcoholism. The mechanisms associated with alcohol and its use (i.e. social–cultural conditions) should be examined in order to understand which emotions dominate (and why) when people choose to engage in this consumable technology. Alcoholism can be classified as a situation in life, while drinking, its surrounding behaviour and antecedents are the actions that lead to this situation. Emotions regarding alcoholism and its long-term consequences, as seen in Korsakoff's syndrome—a preventable memory disorder (Kopelman et al. 2009)—do not rest merely in emotional issues. The control and analysis of consequences are always cognitive (information processing) issues leading to the study of knowledge, otherwise known as hermeneutics (Gadamer 2008) and experience (phenomenology). Thus, differently to human cognitive and emotional encoding, technoethical processes are intimately linked to the whole (Barnes and Thagard 1996; Thagard 2008). In terms of how emotions operate within these cognitive processes, it is useful to refer to research on appraisal theory (Folkman and Lazarus 1984; Frijda 1993). This theory explains benefit–detriment and safety–threat relationships in terms of how emotions emerge through evaluations of encounters and experience at various levels of cognition and through different types of emotions—from basic or primal emotional reactions (Ekman 1999; Ortony and Turner 1990) to higher order cognitive emotions (Clark 2010; LeDouz and Brown 2017).

The basic unit of ethical analysis is the cognitive representation of actions associated with emotional valence. Through evolutionary psychology, humans are programmed to avoid actions that eventually lead to negative emotional states such as harm, pain, humiliation, guilt and/or stress (Neuberg et al. 2011; Petersen et al. 2012). Cognitive analysis renders it possible to understand how humans should act to avoid negative, destructive and exploitation experiences. This type of approach

---

[5]The General Data Protection Regulation (GDPR) begins to influence this practice at the design level through interventions such as cookie consent banners. Yet, arguably from an ethical perspective, more can be done in terms of increasing user awareness of what cookies are and why consent is important.

is useful when designing technology for ethical emotional experience. Difficulties arise, however, due to the fact that ethical experiences and interpretations of ethics differ not only between individuals, communities and cultures but also according to the context. The ethical qualities of AI that are developed to help doctors optimise the effectiveness and efficiency of disease diagnosis and immunisation development, such as needed now with the COVID-19 virus, suggest that such technology is beneficial for a large population of people. However, the scientists, developers and companies that have the resources to develop such powerful systems will also inevitably be able to exploit these findings financially. Thus, the end result may be that the subsequent developments are experienced more as a financial win for those who 'have' (parties who already have strong socio-economic resources) than as a health benefit for the broader global population (a mixture of 'haves' and 'have nots').

Ethical norms can be seen as a consequence of both informal (every day) and formal (or political) discourses and are the subjects of study within the area of discourse ethics (Habermas and Cronin 1993). Thus, discourse is an integral element to consider within technoethical processes. Not only does discourse reveal the semantic and interpretative framing of particular issues within the context of society but also reveals the power relationships, cause and intentionality behind the development of the ethical norms themselves. Hume (1751/1998) neglected to highlight three essential components of ethical generation processes. These components relate to: (1) life circumstances, (2) information analysis and perceived cause–effect relationships and (3) socio-ethical discourse. Firstly, the intertwined relationship between ethics and emotions depends on life circumstances and subjective conditions. That is, assessments of benefit/detriment, winners/losers and ethical interpretation always occur from the standpoint of the individual party and their approach to ethical analysis (their ethical stance: how they use the available information to justify their and others' actions in relation to ethics). Secondly, ethical processing entails the evaluation and analysis of information related to various actions leading to various types of situations, the information produced within the situations and the subsequent outcomes. Thirdly, socio-ethical discourse needs to be incorporated into the analysis in order to define the social and historical properties of a situation. Though Hume theorised the cognitive–affective–behavioural triad in terms of emotions, reason and action, his guillotine irrationally broke the process.

Hume's guillotine is a consequence of a mistaken analysis of the ethical process and the ethicality of actions. He was not concerned with how ethics arise from the dynamics of emotional experience and the simultaneous analysis of situations. Cognitive and emotional aspects of situations are encoded parallel to the evaluation of ethical stance and appraisal. This is why the very question of whether or not (cognitive) facts are utilised to define (emotional) values is senseless. Facts and values are simply two sides of the same mental event. Social discourse operates in a way that generalises ideas, establishing hegemony (Gramsci 2006; Laclau and Mouffe 2014), that relate actions to cognition and subsequent (perhaps idealised) social and individual emotions. Ironically, from a technological development perspective, an accurate analysis of the ethical process as a whole—that incorporates social, biological, political and psychological components—informs the development of weak and strong AI technology, for instance. This is due to the fact that technologists can ascertain the socio-cultural components of emotional experience, from and in conjunction with biological components, meaning that contextually aware emotional systems could be established in relation to information types and surrounding social discourse. But would these systems be ethical? And would this knowledge about the interaction between social knowledge, conventions and moral codes (ethics) and cognitive–affective processing be used for the *right* purposes?

11.6 Conclusion

Knowledge of ethical information processing can help to circumvent Hume's guillotine. Hume makes the fundamental (unsupported) generalised assumption that emotions and cognition are opposites in the human mind. This establishes a binary approach to cognition and emotions, which has led to the popular myth that reason and rationality are based on the cognition of *facts*, whereas emotions and their expression are somewhat fuzzy, subjective *reactions*. However, emotions guide attention, prioritise information, enable and disable memory and shape memory. In so doing, they operate on various levels throughout the society—from the highly individual (subjective) to the mass (social) and political levels. At the beginning and the end of science are always emotions. The very act of separating what is considered factual information and cognition from emotional processing cannot be supported (Dagleish and Power 2004; Thagard 2008). One cannot conclude that the two concepts are opposites based on the fact they are expressed differently. Instead, they are components of the same evolutionary system. This is why it is essential to analyse the ethical uses and emotional consequences of actions when discussing technoethics.

It is not necessary to deem all technologies that are attached to a business model, unethical and controversial to constructive social thought. This is because technology, and how it is supported through economic models, is a result of the world's constructed socio-economic conditions. The problems emerge from an imbalance within benefit–detriment relationships and issues of power and control. Moreover, the intentionality behind the technology development—i.e. produced for diagnosis and world health or for financial gain, destruction and control—shifts the ethical–emotional relationship between controversial and acceptable boundaries. Tobacco, for instance, can be seen to have positive effects in terms of weight control and social health (cigarettes as networking tools). Yet, from the perspective of world health, it causes various types of cancer and lethal cardiac diseases and is still legal and on the global markets due to corporate profit. Nuclear energy is another controversial technology. It is a clean technology and relatively economical to use, yet it is questionable. The consequences of malfunction are so severe, and the destructive effects are so permanent, that it is unclear whether it should be used. The critical evaluative factors involved in tackling this problem are closely linked to ethical questions regarding the risks versus benefits—those who gain and those who potentially lose. The outcomes of such evaluative processes can never be certain. Emotions experienced in relation to these processes will always be conflicting and varying.

There are also destructive technologies. For example, bots spreading intentional disinformation are not meant to increase the quality of human life (Saariluoma and Maksimainen 2012). They are instead intended and designed to influence emotion and shift social discourse for (usually political or financial) gain. They are meant to rupture societies (divide and conquer) and, through instability, to establish power through false structures—e.g. similarly to hackers and individuals guilty of cyber-crime who establish cybersecurity companies. From this perspective, any form of disinformation can only be harmful. As seen in conjunction with the 2016 US pres-idential election and the infiltration of fake news via social media, Soviet propa-ganda similarly hid and manipulated important statistical facts. Thus, disinformation created illusory bubbles, which have been gradually refuted, consequently leading to political changes, mixed emotions and a lack of trust.

These examples show that improving the quality of life, which is ultimately defined emotionally, does not always drive technology design and construction. Human values will always vary. However, technology design is value motivated, rendering value and ethical analysis an important part of modern technology design, particularly where emotions are concerned. The development of ethical norms is grounded in the analysis of emotional situations and vice versa. Thus, ethical process analysis should not end with a seeming understanding of emotions; rather, ethics and emotions should be understood as being intertwined—relying on cognitive–affec-tive, social, cultural, political and contextual interactions and relations. Situations of life are consequences of actions.

Information systems and emerging technology are involved in carrying out increasingly complex actions and processes. It is thus essential to develop ethical capacities to accommodate these systems at the levels of technological logic (input and design), and social costs and benefits, in addition to economic modelling and sustainability. Designing from an ethical perspective means increasing the benefits for as many individuals and parties as possible and being transparent about the cost–benefit dimensions—indicating who gets what, when and how. The operational roles of technology can be very independent; thus, it is essential that they adhere to a code of principles that define ethical practices.

There are two approaches to ethical AI logic: ethically weak and ethically strong AI systems. The former can apply ethical rules that have been programmed into systems, directing behaviour in given situations. In this case, AI would be able to recognise critical features in situations and select actions based on this information. In the latter type of system, ethics are simply a human-implanted feature in recognition- and action-based systems. The outcomes of technoethical processes are sets of guide-lines, referred to here as moral norms, with a codification process that can be termed 'norming'. In this technoethical process—or synthetic ethical–emotional process—moral norms and rules are the outcomes of programming in relation to moral and ethical codes. These inform the norming process, which classifies the combination of situations, actions and contexts resulting in particular circumstances as either positive or negative, depending on the application of these moral norms within the particular situation. From the perspective of a situationally aware AI system, ethical rules are utilised as criteria to evaluate the emotional consequences of deeds. The classic Golden Rule, for example, expresses human experience when people do not take into account how their actions affect others. Thus, by implementing a codified framework that may assess hypothetical alternatives about how the machinery (if conscious) would itself experience the consequences of its own actions, the system may be presented with the right solution regarding what the most appropriate action in those given circumstances would be.

Understanding the origins of social discourse may be considered a 'which came first—the chicken or the egg' type of deed. An individual's primary ethical representations and schemas form the basis of social discourse, while social discourse largely informs an individual's personal ideas about ethics and moral codes. A charismatic figure may influence social discourse and interpretative frames for many communities on numerous subjects. Yet, this charismatic figure would not have been born in a vacuum; rather, they too are the result of social–cultural conditions. Through cultural influence as well as small-scale and large-scale, formal and informal discussions, people form their views about what are the most important and fundamental ethical experiences and respective rules, as examined in scholarship on discourse ethics (Habermas 2009; Habermas and Cronin 1993). In discourse ethics, representations are submitted to argumentative or foundational analysis. Each primary representation or ethical rule is subjected to the foundational discourse. Any ethical rules that cannot be argumentatively supported will be rejected. The discourse itself has layers and sub-discourses. The main outcome is a system of ethical concepts, rules and principles. This chapter describes the unification of emotional, cognitive and social analysis as the ethical information process. It argues that in emotional–technological design, ethical information processing is a key for not only knowing how to design for emotions, or why design teams emotionally experience their technological designs in certain ways, but also important for understanding ethical ways of successfully implementing emotions in AI technology design.

References

Association of Internet Researchers (2020) Ethics. Retrieved 15 May 2020, from: https://aoir.org/ethics/

Austin JL (1975) How to do things with words, 2nd Rev edn. Oxford University Press, London

Ayer AJ (1936) Chapter VI: critiques of ethics and theology. In: Language, truth and logic. Dover Publications, New York. Retrieved 10 May 2020, from: https://ethicsintroduction.weebly.com/uploads/4/4/6/2/44624607/ayer_emotivism.pdf

Bahm AJ (1993) Axiology: the science of values, vol 2. Rodopi, Amsterdam, NL

Barnes A, Thagard P (1996) Emotional decisions. In: Proceedings of the eighteenth annual conference of the cognitive science society. University of California San Diego, pp 426–429

Barnes SB (1996) Internet relationships: the bright and dark sides of cyber-friendship. Telektronikk 92:26–39

Bernal JD (1969) The scientific and industrial revolutions. Penguin, London

Black M (1964) The gap between "is" and "should." Philos Rev 73(2):165–181. https://doi.org/10. 2307/2183334

Borgmann A (2012) The collision of plausibility with reality: lifting the veil of the ethical neutrality of technology. Educ Technol 52(1):40–43

Breslavs GM (2013) Moral emotions, conscience, and cognitive dissonance. Psychol Russ 6(4):65–72

Britannica (2020) The gunpowder revolution c1300–1650. Retrieved 15 May 2020, from: https://www.britannica.com/technology/military-technology/The-gunpowder-revolution-c-1300-1650

Bunge M (1977) Towards a technoethics. Monist 60(1):96–107

Cardon A (2018) Beyond artificial Intelligence: from Human consciousness to artificial consciousness. Wiley, Hoboken, NJ

Cerulo KA (2009) Nonhumans in social interaction. Ann Rev Sociol 35:531–552

Chalmers DJ (1996) Facing up to the problem of consciousness. In: Toward a science of consciousness: the first Tucson discussions and debates. MIT Press, Cambridge, pp 5–28

Chan SY (2006) The Confucian notion of Jing (respect). Philos East West, pp 229–252

Chen A, Treviño LK, Humphrey SE (2020) Ethical champions, emotions, framing, and team ethical decision making. J Appl Psychol 105(3):245–273. https://doi.org/10.1037/apl0000437

Clark JA (2010) Relations of homology between higher cognitive emotions and basic emotions. Biol Philos 25(1):75–94

Cohen JR, Pant L, Sharp DJ (1992) Cultural and socioeconomic constraints on international codes of ethics: lessons from accounting. J Bus Ethics 11:687–700

Cohen JR, Pant LW, Sharp DJ (2001) An examination of differences in ethical decision-making between Canadian business students and accounting professionals. J Bus Ethics 30(4):319–336

Cooley M (1995) The myth of the moral neutrality of technology. AI Soc 9(1):10–17

Dagleish T, Power MJ (2004) Emotion specific and emotion-non-specific components of PTSD: implications for a taxonomy of related psychopathology. Behav Res Ther 42(9):1069–1088

Dennett DC (1993) Consciousness explained. Penguin, London

Derry TK, Williams TI (1960) A short history of technology from the earliest times to AD 1900, vol 231, Courier Corporation

Devillers L (2020) Social and emotional robots: useful artificial intelligence in the absence of consciousness. In: Nordlinger B, Villani C, Rus D (eds) Healthcare and artificial intelligence. Springer, Cham, pp 261–267

Dror Y (1957) Values and the law. Antioch Rev 17(4):440–454

Ekman P (1999) Basic emotions. Handb Cogn Emot 98(16):45–60

Ferrarello S (2015) Husserl's ethics and practical intentionality. Bloomsbury Publishing, London

Findlay JN (1970) Axiological ethics. Macmillan International Higher Education, London

Folkman S, Lazarus RS (1984) Stress, appraisal, and coping. Springer Publishing Company, New York, pp 150–153

Frede D (2017) Plato's ethics: an overview. In: Zalta EN (ed) The stanford encyclopedia of philosophy (Winter 2017 Edition), Retrieved 5 May 2020, from: https://plato.stanford.edu/archives/win2017/entries/plato-ethics/

Frijda NH (1993) The place of appraisal in emotion. Cogn Emot 7(3–4):357–387

Fung YL (1953) A short history of Chinese philosophy. In: Bodde D (Trans.). Macmillan Co, New York, NY

Gadamer HG (2008) Philosophical hermeneutics. University of California Press, Berkeley, CA

Gensler HJ (2013) Ethics and the golden rule. Routledge, London

Giles J (2006) Aboriginal remains head for home. Nature 444(7118):411. https://doi.org/10.1038/444411a

Gramsci A (2006) Hegemony, intellectuals and the state. In: Storey J (ed) Cultural theory and popular culture: a reader, 3rd edn. Pearson, Harlow, pp 85–91

Green SK (2000) The fount of everything just and right? The ten commandments as a source of American law. J Law Relig 14(2):525–558

Habermas J (2009) Europe: the faltering project. Polity, Cambridge

Habermas J, Cronin C (1993) Justification and application: remarks on discourse ethics. MIT Press, Cambridge

Haikonen PO (2020) On artificial intelligence and consciousness. J Artif Intell Conscious 7(01):73–82

Hart SL (1971) Axiology—theory of values. Res 32(1):29–41

Hartman RS (2011) The structure of value: foundations of scientific axiology. Wipf and Stock Publishers, Eugene

Hendrick HW (2009) A sociotechnical systems model of organizational complexity and design and its relation to employee cognitive complexity. In: Proceedings of the human factors and ergonomics society annual meeting, vol 53(16). Sage Publications, Los Angeles, pp 1028–1032

Hiereth H, Prenninger P (2007) Charging the internal combustion engine. Springer Science & Business Media, Berlin/Heidelberg

Hodgson B (1983) Economic science and ethical neutrality: the problem of teleology. J Bus Ethics 2(4):237–253

Hofstede G (1980) Culture's consequences. Sage Publications, Beverly Hills, CA

Hume D (1751/1998) An enquiry concerning the principles of morals. In: Beauchamp TL (ed) Oxford University Press, Oxford

Hume D (1888) Treatise to human nature. Clarendon Press, Oxford

Islamic network groups (2020) First principles of religion: treat others as you would like to be treated (The Golden Rule). Retrieved 15 May 2020, from: https://ing.org/first-principles-religion-the-golden-rule/

Ivanhoe PJ (1990) Reweaving the "one thread" of the analects. Philos East W 40(1):17–33

Jarvenpaa SL, Shaw TR, Staples DS (2004) Toward contextualized theories of trust: the role of trust in global virtual teams. Inform Syst Res 15(3):250–256. https://doi.org/10.1287/isre.1040.0028

Jeffrey C, Dilla W, Weatherholt N (2004) The impact of ethical development and cultural constructs on auditor judgments: a study of auditors in Taiwan. Bus Ethics Q 14(3):553–579

Jonas H (1982) Technology as a subject for ethics. Soc Res 891–898

Joseph CM (2009) Is emotivism more authentic than cognitivism? Some reflections on contemporary research in moral psychology. In: Emotions, ethics, and authenticity, pp 155–178

Kant I (1788/2004) The critique of practical reason. In: Kingsmill T (Trans.). Dover Publishers, Mineola, NY

Kasavin IT (2020) Science and public good: max Weber's ethical implications. Soc Epistemol 34(2):184–196

Knapp S, Handelsman M, Gottlieb M, VandeCreek L (2009) Positive ethics: themes and variations. In: Lopez SJ, Snyder CR (eds) The Oxford handbook of positive psychology. Oxford University Press, Oxford, pp 105–113

Kopelman MD, Thomson AD, Guerrini I, Marshall EJ (2009) The Korsakoff syndrome: clinical aspects, psychology and treatment. Alcohol Alcohol 44(2):148–154

Kron A, Pilkiw M, Banaei J, Goldstein A, Anderson AK (2015) Are valence and arousal separable in emotional experience? Emotion 15(1):35–44. https://doi.org/10.1037/a0038474

Laclau E, Mouffe C (2014) Hegemony and socialist strategy: towards a radical democratic politics. Verso Trade, London

Lazarus RS, Smith CA (1988) Knowledge and appraisal in the cognition—emotion relationship. Cogn Emot 2(4):281–300

LeDoux JE, Brown R (2017) A higher-order theory of emotional consciousness. Proc Natl Acad Sci 114(10):E2016–E2025

MacDorman KF, Green RD, Ho CC, Koch CT (2009) Too real for comfort? Uncanny responses to computer generated faces. Comput Hum Behav 25(3):695–710

Malik K (2014) The quest for a moral compass: a global history of ethics. Atlantic Books Ltd., Camden

Markus HR, Kitayama S (1994) The cultural construction of self and emotion: Implications for social behavior. In: Kitayama S, Markus HR (eds) Emotion and culture: empirical studies of mutual influence. American Psychological Association, Worcester, MA, pp 89–130. https://doi.org/10.1037/10152-003

McGuire M (1977) Mythic rhetoric in *Mein Kampf*: A structuralist critique. Q J Speech 63(1):1–13

McNair LD, Paretti MC (2010) Activity theory, speech acts, and the "'doctrine of infelicity'": connecting language and technology in globally networked learning environments. J Bus Tech Commun 24(3):323–357

Merriam-Webster (2020) Technology. Retrieved 5 May 2020, from: https://www.merriam-webster.com/dictionary/technology

Mill JS (1863) Chapter 2—What utilitarianism is. Utilitarianism. Retrieved 5 May 2020, from: https://www.utilitarianism.com/mill2.htm

Moore GE (1959) Principia ethica, vol 2. Cambridge University Press, Cambridge

Moore GE (1991) The elements of ethics. Temple University Press, Philadelphia, PN

Muraca B (2011) The map of moral significance: a new axiological matrix for environmental ethics. Environ Values 20(3):375–396

Neuberg SL, Kenrick DT, Schaller M (2011) Human threat management systems: self-protection and disease avoidance. Neurosci Biobehav Rev 35(4):1042–1051

Newell AS, Simon HA (1972) Human problem solving. Prentice-Hall, Englewood Cliffs, NJ

Norman DA (1976) Memory and attention: an introduction to human information processing, 2nd edn. Wiley, Hoboken, NJ

O'Shaughnessy N (2009) Selling Hitler: propaganda and the Nazi brand. J Public Aff Int J 9(1):55–76

Oatley K (1992) Best laid schemes: the psychology of the emotions. Cambridge University Press, Cambridge

Ortony A, Turner TJ (1990) What's basic about basic emotions? Psychol Rev 97(3):315–331. https://doi.org/10.1037/0033-295x.97.3.315

Petersen MB, Sznycer D, Cosmides L, Tooby J (2012) Who deserves help? Evolutionary psychology, social emotions, and public opinion about welfare. Political Psychol 33(3):395–418

Prince JD (1904) The code of Hammurabi. Am J Theol 8(3):601–609. Retrieved 5 May 2020, from: https://www.journals.uchicago.edu/doi/pdfplus/10.1086/478479

Ramsey FP (2016) Truth and probability. Readings in formal epistemology. Springer, Cham, pp 21–45

Robinson MD, Storbeck J, Meier BP, Kirkeby BS (2004) Watch out! That could be dangerous: valence-arousal interactions in evaluative processing. Pers Soc Psychol Bull 30(11):1472–1484

Rousi R (2013a) From cute to content: user experience from a cognitive semiotic perspective. In: Jyväskylä studies in computing. University of Jyväskylä Press, Jyväskylä, p 171

Rousi R (2013b) The experience of no experience: elevator UX and the role of unconscious experience. In: Proceedings of international conference on making sense of converging media. ACM, New York, NY, pp 289–292

Rousi R (2014) Unremarkable experiences: designing the user experience of elevators. Swed Des Res J 11:47–54

Rousi R, Saariluoma P, Leikas J (2010) Mental contents in user experience. In: Proceedings of MSE2010 V. II 2010 international conference on management and engineering, 17–18 Oct 2010. Wuhan, China, pp 204–206

Rumelhart DE, Lindsay PH, Norman DA (1972) A process model for long-term memory. In: Tulving E, Donaldson W (eds) Organization of memory. Academic Press, Cambridge, MS

Saariluoma P (2003) Apperception, content-based psychology and design. In: Lindemann U (ed) Human behaviour in design. Springer, Berlin/Heidelberg, pp 72–78

Saariluoma P (2020) Hume's guillotine in designing ethically intelligent technologies. In: Taiar R, Gremeaux-Bader V, Aminian K (eds) International conference on human interaction and emerging technologies. Springer, Cham, pp 10–15

Saariluoma P, Leikas J (2020) Designing ethical AI in the shadow of Hume's guillotine. In: Ahram T, Karwowski W, Vergnano A, Leali F, Taiar R (eds) International conference on intelligent human systems integration. Springer, Cham, pp 594–599

Saariluoma P, Maksimainen J (2012) Intentional disinformation and freedom of expression. Int Rev Soc Sci Humanit 3(2):9–20

Saariluoma P, Rousi R (2015) Symbolic interactions: towards a cognitive scientific theory of meaning in human technology interaction. J Adv Humanit 3(3):310–324. Retrieved 10 May 2020, from: file://fileservices.ad.jyu.fi/homes/rerousi/Downloads/SYMBOLIC_ INTERACTIONSTO-WARDS_A_COGNITIVE_SCIENTIFI.pdf

Shotter J (1995) In conversation: joint action, shared intentionality and ethics. Theor Psychol 5(1):49–73

Stevenson CL (1944) Ethics and language. Yale University Press, New Haven, NJ

Stone J, Cooper J (2001) A self-standards model of cognitive dissonance. J Exp Soc Psychol 37(3):228–243

Tegmark M (2017) Life 3.0: being human in the age of artificial intelligence. Knopf, New York City, NY

Thagard P (2008) Hot thought: mechanisms and applications of emotional cognition. MIT Press, Cambridge, MA

Turkle S (2007) Authenticity in the age of digital companions. Interact Stud 8(3):501–517

Turnbull P (2007) British anatomists, phrenologists and the construction of the Aboriginal race, c. 1790–1830. Hist Compass 5(1):26–50

United Nations (2020) Some articles from the UN's universal declaration of human rights. Retrieved 15 May 2020, from: https://news.un.org/en/tags/universal-declaration-human-rights

van Roojen M (2018) Moral cognitivism vs. non-cognitivism. In: Zalta EN (ed) The stanford encyclopedia of philosophy (Fall 2018 Edition), Retrieved 10 May 2020, from: https://plato.stanford.edu/archives/fall2018/entries/moral-cognitivism

von Wright G (1963) The varieties of goodness. Routledge and Kegan Paul, London

Westermarck E (2017) Ethical relativity. Routledge, London

Western Australian Museum (2017) '67 Referendum—WA 50 years on: dispelling myths. Retrieved 15 May 2020, from: https://museum.wa.gov.au/referendum-1967/dispelling-myths

Winkielman P, Berridge KC (2004) Unconscious emotion. Curr Dir Psychol Sci 13(3):120–123

Wolff T (2019) That's close enough: the unfinished history of emotivism in close reading. PMLA 134(1):51–65

Zardiashvili L, Fosch-Villaronga E (2020) "Oh, Dignity too?" said the robot: human dignity as the basis for the governance of robotics. Mind Mach 30:121–143. https://doi.org/10.1007/s11023-019-09514-6

Zerubavel E (1982) The standardization of time: a sociohistorical perspective. Am J Sociol 88(1):1–23

Printed in the United States
by Baker & Taylor Publisher Services